CHIMPANZEE MEMOIRS

CHIMPANZEE MEMOIRS

STORIES OF STUDYING AND SAVING OUR CLOSEST LIVING RELATIVES

EDITED BY

STEPHEN ROSS AND LYDIA HOPPER

ILLUSTRATIONS BY DAWN SCHUERMAN

Columbia University Press   *New York*

Columbia University Press
Publishers Since 1893
New York   Chichester, West Sussex
cup.columbia.edu

Library of Congress Cataloging-in-Publication Data
Names: Ross, Stephen R., editor. | Hopper, Lydia M., editor.
Title: Chimpanzee memoirs : stories of studying and saving our
closest living relatives / edited by Stephen Ross and Lydia Hopper ;
illustrations by Dawn Schuerman
Description: First edition. | New York : Columbia University Press,
[2022] | Includes bibliographical references.
Identifiers: LCCN 2021038565 (print) | LCCN 2021038566
(ebook) | ISBN 9780231199285 (hardback) | ISBN 9780231199292
(trade paperback) | ISBN 9780231553032 (ebook)
Subjects: LCSH: Primatologists—Biography. |
Chimpanzees—Behavior—Research. | Chimpanzees—Research. |
Chimpanzees—Conservation.
Classification: LCC QL26 .C45 2022 (print) | LCC QL26 (ebook) |
DDC 591.092/2 [B]—dc23
LC record available at https://lccn.loc.gov/2021038565
LC ebook record available at https://lccn.loc.gov/2021038566

Cover design: Julia Kushnirsky
Cover illustration: Sarah Wilkins

CONTENTS

Foreword *vii*

Acknowledgments *xv*

1 Jane Goodall 1

2 Lilly Ajarova 13

3 Richard Wrangham 24

4 John Mitani 37

5 Caroline Asiimwe 50

6 Anne Pusey 58

7 Tetsuro Matsuzawa 70

8 Christophe Boesch 80

9 Andrew Whiten 91

10 Melissa Emery Thompson 100

11 David Koni 114

12 Tatyana Humle 128

13 Brian Hare 142

14 Raven Jackson-Jewett 151

15 Frans de Waal 163

16 Elizabeth Lonsdorf 175

Afterword *187*

Suggested Reading *193*

FOREWORD

STEPHEN ROSS

Dr. Stephen Ross is the director of the Lester E. Fisher Center for the Study and Conservation of Apes at Lincoln Park Zoo in Chicago and coeditor of Chimpanzee Memoirs, Chimpanzees in Context, *and* The Mind of the Chimpanzee. *He has studied chimpanzee behavior, cognition, and welfare for over twenty-five years working in lab, zoo, and sanctuary settings. He has served as chair of both the Chimpanzee Species Survival Plan (SSP) and the board of directors at Chimp Haven; he also founded Project ChimpCARE in 2009.*

ORIGINS

We all love a good origin story. And the best origin stories are of course from those that become superheroes, such as Spiderman's fateful encounter with a radioactive spider or a young Bruce Wayne witnessing the murder of his parents, which eventually created his transformation into Batman. Such fanciful comic book tales are inspiring because they demonstrate how anybody, following a twist of fate or seemingly innocuous experience, can grow to become a transformative figure to others.

The psychologist Robin Rosenberg wrote that the importance of superhero origin stories are not that they show us how to become super, but how to be *heroes*, choosing altruism and the greater good over more materialistic pursuits. She describes the transformation of "ordinary people" into heroes as undergoing three relatable experiences: trauma, destiny, and sheer chance, themes you will hear repeated in this collection of essays. These familiar trajectories inspire us and provide us with models of overcoming adversity and discovering inner strength with which to do good in the world.

Not all of us aspire to fight crime, and frankly, the best origin stories are those from real life. As a youngster growing up in Canada in the 1980s and 1990s, I had my own set of heroes. I am so fortunate to count my parents among them, but also athletes like Roberto Clemente and actors like Harrison Ford. But as I grew increasingly interested in primates and especially chimpanzees, I became aware that even the field of primatology had heroes.

Over the past twenty-five years I have worked in this field, it has become increasingly clear that I am not the only one with a piqued interest in knowing more about the people who have dedicated their lives to studying, understanding, and protecting our closest living relative, the chimpanzee. In a way, these heroes are even more remarkable than the comic book characters fighting for humanity because they are fighting not for their own species but for another!

The origin story of this book itself is interesting. My coeditor, Lydia Hopper, and I had just finished hosting an international meeting of chimpanzee experts entitled *Chimpanzees in Context*. It was the fourth in a long-running series of such chimp-centric meetings held every decade and was a wonderful opportunity to interface with some of the greatest chimpanzee scientists in the world. As we wrapped up the organization for the event, two things became clear to us. The first is that we really were on the precipice of a new era in chimpanzee science. Some of the real luminaries in the field were either retired or preparing to do so, and a whole new generation of bright minds was emerging in the field. Capturing this moment, the passing of the proverbial baton, seemed too good to pass up. Second, despite the vast collection of books about chimpanzees, there seemed relatively little in the way of coverage of the people who conducted such work. As such, this book is not about chimpanzees but about the people who study them and work to protect them around the world.

The collection of essays we've gathered here tell the stories of these heroes: what inspired them, what shaped their career choices, and what motivates them to continue to find solutions for the many challenges that chimpanzees face today. Some of these heroes you know and others you don't. They are both junior and senior in their careers, and their work spans anthropology, psychology, ecology, conservation, biology, and environmental ethics. These are the stories of people growing up in the English countryside, the suburbs of Tokyo, and the villages in Congo. Of their first exposures to chimpanzees in a rustic zoo, a New York elevator, and in the forests surrounding their homes. And how they overcame unfair limitations placed on them by others because of their education, class, gender, and race. Sixteen chapters and sixteen stories answer the question that so many people have asked us: "How did you get to work with chimpanzees?"

DIFFERENT PATHS

A book about those who have dedicated their lives to chimpanzees could start no other way than with Dr. Jane Goodall. Of course, her name is virtually synonymous with the study of this species, and in many ways, she is the uber-hero of the community, having inspired the work of most of the other scientists voicing their stories in this book. Her beginnings are known to not only primatologists but a much broader swath of the world's population. Carrying not much more than the confidence placed in her by Louis Leakey and her mother, Jane's path from a university secretary to global conservation icon inspired so many other authors in this collection.

In fact, Jane's legacy is overtly on display here, with Jane's direct protégé Anne Pusey sharing her stories of meeting Jane and being thrust into the fieldwork at Gombe Research Centre in the 1990s. And, subsequently, Anne's student Elizabeth Lonsdorf represents the third generation of influential scientists to come out of this academic lineage.

But Jane is certainly not the only well-established chimpanzee expert with decades of field experience under their proverbial belt. Richard Wrangham, Tetsuro Matsuzawa, John Mitani, and Christophe Boesch have each studied chimpanzees for at least twenty years and collectively represent some of the most influential thinking about wild chimpanzee behavior, ecology, and cognition. Theirs is a legacy of a generation of intensive groundbreaking fieldwork that sets the bar for our understanding of wild chimpanzees and continues to produce the next generation of scientists dedicated to that topic, including two others represented in this book: Melissa Emery Thompson and Tatyana Humle.

While much public attention has been given to Western scientists crossing oceans to study chimpanzees in Africa, incredible work has long been accomplished in the very countries in which wild chimpanzees live. David Koni grew up in the Republic of the Congo and moved from the capital city of Brazzaville to a small town in the north. There he discovered the world of chimpanzees firsthand as a child visiting his uncle and began a journey that has led him to participate in critical conservation work protecting chimpanzees in his home country. Likewise, the story of Caroline Asiimwe is a remarkable tale of perseverance and compassion that started at the Uganda Wildlife Conservation Education Centre

("Entebbe Zoo") and continues today high along the Albertine Rift at the Budongo Conservation Field Station.

Whether rescuing orphaned chimpanzees in Africa or providing a forever home for former laboratory subjects in the United States, sanctuaries are playing an ever-increasing role in caring for displaced chimpanzees. Lilly Ajarova established herself as a pioneer in her native Uganda, rising to lead the Ngamba Island Chimpanzee Sanctuary. Raven Jackson-Jewett, the head veterinarian at Chimp Haven, the world's largest chimpanzee sanctuary, shares her experiences growing up in New York City and the challenges she faced overcoming the limitations that people unwisely set on her. And Brian Hare, among the first to recognize the vast research potential of African sanctuaries, weaves stories contrasting his experiences with chimpanzees and bonobos, putting much of what we know about chimpanzees into context with their sister species.

Though much of the most iconic chimpanzee work has taken place in forests across equatorial Africa, such work has been complemented and enhanced by the behavior and cognition work taking place in captive settings. The work of Frans de Waal is well known to science-savvy audiences around the world through his best-selling books, but here he relates how he bucked traditional sociological paradigms in his native Netherlands and used the study of chimpanzee social lives to expand our understanding of the origins of human behavior. Likewise, Andy Whiten shares how using artificial fruits resulted in his seminal work, which explains the broad scope of chimpanzee culture and traditions that we know so well today.

CONVERGENCE

For many reasons, this book is personal to me. Though I have been studying chimpanzees in a variety of settings for a quarter-century,

I still remember clearly that feeling of wonder when reading about Jane's first moments at Gombe in a ragged copy of *National Geographic* and that spark of intense curiosity when I immersed myself in de Waal's seminal book, *Chimpanzee Politics*. And the deep admiration I felt when, as an early career scientist myself, I first met Tetsuro Matsuzawa in a crowded train station in Kyoto.

These scientists, like all the authors in this book, have taken very different and very unique paths to get to where they are today. The stories they tell here show that they are just like you and me, facing challenges and overcoming them to converge on a common goal: to understand and protect our closest living relatives on this planet. For those aspiring to follow a similar path, these stories may serve as a guide and as inspiration. But we must also remember what serves as inspiration for them—the chimpanzees themselves.

You will read how chimpanzees have inspired our authors and how specific individuals have influenced them in ways they may not have expected. For me, it was a chimpanzee named Drew at Yerkes National Primate Research Center in Atlanta. When I met Drew, I was just starting my career with chimpanzees, but only a few years previously, I had spent much time dreaming of that opportunity. For me, chimpanzees represented that amazing dichotomy of being so remarkably similar to us and yet so different, so primitive and yet so complex, so savage and so gentle. Drew was all of that and getting to know him as an individual with specific needs, preferences, and moods was a turning point in my understanding of chimpanzees. Throughout the book, you will see illustrations of other such chimpanzees that have had made similarly powerful impacts. While we may hold up Jane Goodall and others as heroes, for many of the authors here the chimpanzees themselves are imbued with a superpower. The chimpanzees inspire and motivate us to understand, study, and protect them for future generations to witness.

Drew

These are the memoirs of the heroes who have dedicated their lives to chimpanzees, the chimpanzees that inspired them, and the life paths that converged to make these stories happen.

ACKNOWLEDGMENTS

L ike the chimpanzees we strive to understand and protect, we humans are a social species, so it's not surprising that this book would not have been possible without the critical support of friends, colleagues, and family.

We are so appreciative of the sixteen expert contributors, many of them on the verge of peaceful retirement, who responded to our most unusual request to recount the early days of their lives and careers. We asked all the authors to be reflective and introspective, and the results were very rewarding to say the least.

The book comes directly from our task of organizing *Chimpanzees in Context*, the fourth in the Understanding Chimpanzees symposia series; we are grateful to everyone who helped us pull that event together in 2016. Furthermore, the idea for a book that was very different to traditional academic outputs was the result of a conversation with Christie Henry, who was then the editorial director at University of Chicago Press. She had the foresight to convince us that the benefits of producing not one but *two* books would be well worth the effort it would take to produce them. Thus, this book represents a sister volume to *Chimpanzees in Context*, which was published in 2020. We also want to thank Miranda Martin, our editor at Columbia University Press, who has been our advocate on this project since day one and has guided us as we compiled and edited this volume.

The raw content that resulted in these chapters came in many forms, from informal communications, essays, written responses to interview questions, and recorded conversations with our contributors. We are so thankful to Jillian Braun, who conducted multiple long interviews over the phone with several contributors and to Jocelyn Woods and the other research interns who helped collate much of the material in an organized way, making our editorial work so much easier. We are also appreciative of Dawn Schuerman, who provided the beautiful illustrations of some of the individual chimpanzees who inspired our authors.

Of course, the backbone to all these efforts is the support of the people who encouraged and motivated us, even when the task seemed overly daunting. We are very fortunate to work with our amazing colleagues at Lincoln Park Zoo, especially our dedicated team at the Lester E. Fisher Center for the Study and Conservation of Apes. Their help has been so important in keeping our research and writing efforts going, especially during this difficult past year with the COVID-19 pandemic. In particular, we thank David Morgan, Christina Doelling, Katie Cronin, Marisa Shender, Jesse Leinwand, Sarah Huskisson, and Ben Lake. Additionally, the Lincoln Park Zoo administration has been unrelentingly supportive of our requests to take on these writing projects. And the list of friends and family that not only put up with our project juggling but encourages us to take on new challenges is so very appreciated. Thank you in particular to Megan Reinertsen Ross and Andrew Steets and, of course, to our parents in Canada and the UK, who have supported our passion for our careers working with chimpanzees.

CHIMPANZEE
MEMOIRS

1

JANE GOODALL

Dr. Jane Goodall, DBE, is universally known for her pioneering studies of the chimpanzees of Gombe National Park in Tanzania and her lifetime of advocacy work. Though some of her early research is now sixty years old, her work at Gombe remains extremely influential. At only thirty-five square kilometers, Gombe is one of the smallest national parks in Tanzania. It lies along the hills of the eastern shore of Lake Tanganyika and today is home to fewer than one hundred well-studied chimpanzees that continue to be observed and protected by researchers.

DREAMING OF AFRICA

At age ten, I read Edgar Rice Burroughs's *Tarzan of the Apes* and resolved that I would go to Africa, live with wild animals, and write books about them. I had no thought of becoming a scientist—just a naturalist, an explorer. Almost everyone laughed at me. After all, we had very little money. Africa was so far off, and we knew little about it. World War II was raging. And I was *just a girl.* "Dream about something you can achieve," they said. But I was fortunate to have a supportive mother who told me, "You will have to work hard, take advantage of opportunities, and if you never give up you may find a way."

I never dreamed of living with animals as absolutely exotic as chimpanzees. I would have in fact studied *any* animal in Africa— I just wanted to be out in the wild. It was after I had saved up the fare to go stay with a friend that I met Louis Leakey, and he was the one who suggested that I should work with chimpanzees. He was impressed by how much I knew about animals. He was not at all concerned that I had not been to college—he wanted someone whose mind was uncluttered by the very reductionist thinking of the ethologists of the time. And he felt that women might be more patient out in the field. Like my mother, he had faith in me. He did not try to teach me what to do—there was no protocol for me to follow. And so I began my research with only my instincts and common sense to guide me. Little did we think that I was beginning a study that would last for more than sixty years. What a journey it has been!

THOSE FIRST DAYS IN THE FOREST

The first time I went to Africa, in 1957, I was twenty-three. In those days it was almost of unheard of for a young woman to travel abroad on her own. I went by boat, which was the normal way of travel in those days. It was during that trip that I met Louis Leakey and worked for him as a secretary in the Natural History Museum. Then I had to wait, back in England, until he had found money for what seemed to almost everyone a crazy undertaking. In fact, the British authorities in what was then Tanganyika, a last outpost of the crumbling British Empire, refused to allow me to go to Gombe on my own—and it was my wonderful mother who volunteered to accompany me. She was a perfect companion and helped me in many ways.

While I was out in the forest from dawn to dusk, she stayed in our little camp, and she soon started a "clinic," handing out aspirins, Epsom salts, and other simple medicines to the fishermen who camped along the beach in the fishing season. Thus, from the very start she helped build up an excellent relationship with the local community. But most importantly, from my point of view, she helped to boost my morale when I returned to camp, weary and depressed. For in the early days, if I managed to get anywhere near the chimpanzees, they would quickly vanish into the forest, highly suspicious of this strange white ape who had suddenly invaded their world. And as days became weeks and weeks became months, I became increasingly depressed. I knew I could gain the chimps' trust in the end—but I only had money to stay for six months and there would be no more unless I got results. But when I got back in each evening, dejected, there was Mum to greet me, and while we sat by our little fire and had the supper prepared by our cook, Dominic—mostly from tins—she would boost my morale by pointing out that, in fact, I was learning quite a lot, albeit from a distance, through binoculars. I had found a peak from which I could overlook two valleys. I learned what foods they were eating—I would go and collect specimens after they had gone and my mother would press them so that they could be subsequently identified. I saw how they sometimes traveled alone, sometimes in different-sized groups. And I was also learning about their different calls.

It was sad that I made my first really significant observation just two weeks *after* my mother had to return to England. It was during the third month when I saw David Greybeard, the first chimpanzee who had begun to lose his fear of me, using a grass stem to fish for termites and stripping leaves from a twig, creating another tool for the same purpose. I knew this would really thrill Leakey as only humans were thought to use and make tools. And I so missed being able to share my excitement with my mother.

THE CHIMPANZEES WHO SHAPED ME

During the years when I spent almost all my time at Gombe, I got to know the chimpanzees of the Kasakela community as well as I had known my friends at school. Each one had his or her own personality and they were as different, one from the other, as we are. David Greybeard was very special and not only because he demonstrated a use of tools. He was also the one that taught me chimpanzees eat meat. And from him I learned that the top-ranking male, the alpha, is not necessarily a *leader* per se. David's close friend, Goliath, was the alpha in the early 1960s and, with his dramatic charging displays and courageous personality, commanded the respect of the other chimpanzees, but it was David, with his gentle but determined character, who was so often followed by the others when he set off to a new feeding place. He was quick to reach out and gently touch or embrace those that approached and asked for reassurance. Youngsters, in particular, ran to him for comfort if their mothers were not nearby. And on one memorable occasion he even reassured me. I held out a palm nut in my hand. He obviously did not want it. But looking into my eyes he reached out, took and dropped the nut, then gave my hand a reassuring squeeze. He knew that my motive was good—and I understood that he knew.

Mike was a small and very low-ranking male in the dominance hierarchy under Goliath's rule, and he taught me how it is possible to get to the top simply by using intelligence. He learned to incorporate up to three empty kerosene cans, found in my camp, to enhance his charging displays. Not surprisingly, the other males rushed out of the way—and he got to the top after only a few months.

Then there were the three females, Flo, Passion, and Patti, who taught me so much about chimpanzee mothering and, moreover, that there are good and not-so-good mothers. Flo was an excellent

David Greybeard

mother, protective but not overly so, and also affectionate and playful. Above all, like my own mother, Flo was supportive. When I first got to know her, she must have been well over forty, but even with her teeth worn to the gums, and her hair thinning, she would fearlessly charge an adult male baboon if he threatened any of her offspring. She would even go to help her adolescent son, Faben. Passion was the absolute opposite, seldom waiting for her infant, Pom, to climb onto her back before moving off. She hardly ever played with her, and it was up to Pom to keep out of harm's way, for Passion would not often go to her aid. Interestingly, she was more attentive to her subsequent infant. And Patti was so utterly incompetent that she did not even know how to carry her first infant as she traveled and was apt to place her hand under his rump, so his head bumped along the ground. Not surprisingly, he did not live long—and her second infant only survived in spite of Patti's poor maternal skills. It was only when she had her third baby that she started to show reasonably appropriate maternal behavior. Both Passion and Patti taught me that while some mothering skills are clearly innate, experience also plays a critical role in the development of appropriate maternal behavior. Over time I also learned that the practice a juvenile daughter gets as she plays with and carries an infant sibling also plays a role in developing her subsequent maternal behavior.

MY DAMASCUS MOMENT

In 1986, I helped to organize a conference, "Understanding Chimpanzees," in Chicago. For this meeting, my friend Paul Heltne and I brought together, for the first time, scientists studying chimpanzees in the field and some who conducted (noninvasive) research on captive chimpanzees. The meeting consisted mostly of scientific talks comparing chimpanzee behavior in different parts of

Africa, but we also had a session on conservation and another on the poor conditions in some captive situations—bad zoos, the cruel training of chimpanzees used in entertainment, and the use of our closest living relatives in medical research laboratories. There was a secretly filmed video of infant chimpanzees in incubators—containers that looked like microwave ovens with just one small slit in the steel walls and air going in through ventilators. I could not sleep that night.

That conference was what I call my "Damascus moment." I arrived in Chicago as a scientist, planning to spend forever in the forests I loved learning about the chimpanzees. I left Chicago as an activist. I had to visit some of the labs to see the conditions with my own eyes before beginning a long battle, helped by many, to end invasive research on chimpanzees. And I had to travel around Africa to see the problems faced by wild chimpanzees in order to try to do something to help. It was the start of more or less nonstop traveling around the world giving lectures, going to meetings, and raising awareness of what was going on. And during these years I gradually learned more and more about the harm we humans are inflicting on animals and the natural world. The destruction of forests, woodlands, wetlands—just about every habitat. The pollution of the air as a result of fossil-fuel emissions. The pollution of the water caused by industrial, agricultural, and household waste, much of which ends up in the oceans. We face a climate crisis, caused by the greenhouse gases (especially carbon dioxide, methane, and nitrous oxide) that circle the globe like a blanket, trapping the heat of the sun. This is causing the melting of the arctic ice, the glaciers, and the permafrost (which releases a great deal of methane). Sea levels rise as the water heats and ice melts. And everywhere weather patterns are changing: hurricanes are worse, flooding is more severe, and droughts are leading to terrible wildfires around the world.

As a result of all these changes, there is a loss of biodiversity around the world; entire species are vanishing forever—in fact, we are in the midst of the sixth great extinction. And it is caused not only by habitat destruction but also by the bushmeat trade—the commercial hunting of wild animals, trophy hunting, and the trafficking of animals and their body parts around the world for food, medicine, and exotic pets.

I have also learned more and more about the vast extent of human suffering—war, genocide, domestic violence, poverty, discrimination, drugs, and crime. The desperate plight of refugees and migrants, fleeing violence or worsening conditions caused by climate change.

THE POWER OF EMPATHY AND
THE POLITICS OF SCIENCE

When I arrived from the field to start work on my PhD at Cambridge University, I had never even been to university and had no bachelor's degree. I was nervous—and shocked when some of the professors told me that I had done my study all wrong. I should not have named the chimpanzees but given them numbers. And I could not talk about chimpanzees having *personalities*, or *minds capable of solving problems*, and I should absolutely not attribute *emotional states* such as joy, sadness, despair, grief, and so on to any animal. That was the height of anthropomorphism, a cardinal sin of ethology. All those things were unique to humans, I was told. Fortunately, I had learned very early on that this was not true—from my dog, Rusty. Anyone who has shared their life in a meaningful way with any animal companion knows that of course they have personalities. Of course they have emotions. Of course they have minds—we are part of and not separate from the animal kingdom. But back then, in the mid-1960s, it was taught that there was a difference of *kind* between us and other animals.

I was told repeatedly that empathy has no place in science and that a scientist must at all times be objective; this is not possible if you have any feelings of empathy. In fact, this is not true. Empathy can help you understand, intuitively, the motivation behind certain behavior. This can then serve as a basis for scientific investigation—is my intuition true or false? In fact, in my opinion, it is a lack of empathy that has enabled scientists to carry out many very cruel experiments on sentient beings.

I was also shocked at the response of some scientists when I wrote about the intercommunity violence that I saw at Gombe. I described how male chimpanzees of one community attacked and killed individuals from a neighboring community. I described that as *in-group* and *out-group* behavior and likened it to a form of primitive warfare. That was the time (in the early 1970s) when there were bitter debates about whether human aggression was innate or whether it was learned. I was told I should not dwell on the violent behavior of the chimpanzees as it provided support for those who believed human aggression was innate and that therefore war was inevitable. It was my first introduction to the politics of science.

Gradually, things have changed. It was because chimpanzees are so like us biologically—we differ genetically by only just over one percent, and there are striking similarities in our immune systems and the composition of the blood—that it was believed that they would be ideal models for medical research. The similarities in psychology and social behavior—the fact that they could feel fear, frustration, boredom, and pain—were seemingly considered irrelevant. But as it became more generally recognized that chimpanzees, along with so many other animals, are sentient beings, a growing group of people, including many scientists, began working to bring invasive scientific research on captive chimpanzees to an end.

I was fortunate to meet Francis Collins very soon after he was appointed director of the National Institutes of Health (NIH) in the United States. Our conversations included the scientific and ethical aspects of ending invasive chimpanzee research. All of this culminated in 2015 when the task force appointed by Francis concluded that the research being conducted on federally owned chimpanzees was not beneficial to human health advances. With great courage he decided to end NIH invasive research with chimpanzees and began moving all the federally owned laboratory chimpanzees to Chimp Haven, a sanctuary in Louisiana. I have visited Chimp Haven, and it is a beautiful, peaceful place for these chimpanzees to live out their lives after having served as research subjects for so long.

WHAT WILD CHIMPANZEES FACE

There are probably fewer than three hundred thousand chimpanzees still living wild in Africa today. This number is just a small fraction of the population levels that were present historically, and this loss is because of threats that chimpanzees continue to face and which continue to grow.

In some areas, the destruction of their habitat is driving chimpanzees to extinction. This may be caused by the growth of human populations, and their livestock, which gradually encroach further and further into chimpanzee habitats. Or by commercial logging and mining operations. In some areas of Africa, forests are cleared for oil palm plantations—although fortunately this has not become the nightmare situation that is destroying the rainforests of Asia. Then there is the bushmeat trade—the *commercial* hunting of wild animals for food (chimpanzees are prized meat in some African cultures while in others they are never eaten). Some chimpanzees are caught in wire snares set by hunters for animals such as antelopes, bush pigs, and so on, and while chimpanzees may be able to break the wire, they cannot loosen the noose that becomes tighter and tighter around a wrist or ankle. Where we are working in Uganda, many individuals lost a hand or foot as a result of being snared and some sadly died afterward from gangrene. Then there is the live animal trade—chimpanzee mothers being killed so that their infants can be more easily stolen and then sold. Some are exported illegally to other countries as pets, to disreputable zoos, or for entertainment in circuses and films. As a result of the bushmeat trade, a good many orphaned infants are sold in African markets as pets. Furthermore, as chimpanzees are forced to live ever closer to humans, they are at risk of catching human infectious diseases. At Gombe, there was a terrible epidemic of polio. Chimpanzees are especially vulnerable to respiratory diseases, and that is why the COVID-19 pandemic is of great concern to all of us working in the field or caring for orphans in sanctuaries.

Thus, the list of threats facing chimpanzees—and many other species—is very long and getting longer, which is why it is desperately urgent that we get together to tackle the different problems.

And we cannot hope to save them unless we work with local communities, helping them find ways of making a living that do not involve destroying their environment. For one thing, we should work to alleviate poverty—for when you are very poor, and your population is growing, you will cut down the last trees in order to get new land for growing food or making charcoal for money. The Jane Goodall Institute's TACARE ("Take Care") program, which I began in 1994, works with people in 104 villages throughout the remaining habitat of Tanzania's chimpanzees. The program is also growing in six other countries where the Institute works with chimpanzees. The people have become or are becoming our partners in conservation.

LOOKING TO THE FUTURE

All our efforts to conserve chimpanzees and their forests are of little use if new generations do not grow up to be better stewards of the planet than we have been. I began Roots & Shoots with twelve high-school students in Tanzania in 1991. The movement is now in eighty-six countries and growing. There are now hundreds of thousands of active groups for all ages from kindergarten all the way through university. Each group chooses and tackles projects in three areas—to make the world better for other people, for animals, and for the environment. And there are now thousands of Roots & Shoots alumni, those who participated in the movement in school or college. And all those I have met retain the values they acquired as members—knowing we make a difference every day, choosing what sort of difference we make, and respecting each other and the natural world and animals. When young people understand the problems and are empowered to act, their dedication and energy is amazing. They *are* changing the world!

The "Understanding Chimpanzee" conferences have continued to take place every ten years in Chicago, and I have attended each one of them. The last two have been hosted by Lincoln Park Zoo; the scientists there, who have edited this book, have been instrumental in keeping this important tradition going so new generations can learn about all the threats chimpanzees face and what we must do to cope with those challenges. And each conference sees yet more energized and dedicated individuals who are studying or working to conserve and improve the well-being of chimpanzees.

Many of these chimpanzee researchers are telling their stories in this book. There is so much hard work still required to save chimpanzees and to save this planet. So many challenges to face and hurdles to overcome. But as my mother told me as we sat around the fire at Gombe, tired from a day in the forest, we have to have hope. The hope that tomorrow will bring change. The hope that we shall all do our bit to make this a better world for all.

2

LILLY AJAROVA

Lilly Ajarova is the chief executive officer of the Uganda Tourism Board, but for almost fifteen years she served as executive director of the Chimpanzee Sanctuary and Wildlife Conservation Trust, which operates Ngamba Island Chimpanzee Sanctuary. The sanctuary is a tranquil island of hundred acres in Lake Victoria, Uganda, and is home to around fifty chimpanzees rescued from across East Africa. Here, chimpanzees live a peaceful existence in huge, forested habitats, free from the perils of the illegal bushmeat trade and destructive deforestation activities. Lilly's leadership at Ngamba was instrumental in growing the sanctuary's reputation and capacity, and today she is widely recognized as one of Uganda's conservation champions.

AN AWAKENING

Every time I return to Ngamba Island Chimpanzee Sanctuary, something in me is awakened. A feeling of nostalgia washes over me as I watch the chimpanzees go on with their daily lives. Most of them I know very well, and I remember how much work has gone into teaching them that the humans at the sanctuary are not a threat to them. Helping them learn to trust people has been an important part of my work over the years, and there is a deep satisfaction in this process that is difficult to explain. To best understand it, I must return to my childhood.

I grew up in Angal, a small village in northwestern Uganda. My parents worked in the missionary hospital there—my dad as a doctor and my mother as a nurse. By the age of five, I was already going to the hospital to help my parents by helping feed the malnourished children and dispensing medicine. I have fond memories of my dad taking me on a bumpy ride on his Yamaha motorbike to visit patients in neighboring villages. My dad was a naturalist, and he always took me for long hikes and visits to the national parks whenever he could spare the time. I can still vividly recall my first encounter with a charging elephant in Murchison Falls National Park when I was just six years old! Later in life, I attended Makerere University to study psychology and sociology for my undergraduate degree and then traveled to Austria to study at the International College of Tourism and Management in Krems. When I returned to Africa, I completed my master's degree in business administration and then got a job with a tour and travel company. I put together travel itineraries for tourists visiting Uganda and occasionally helped guide them to its many beautiful national parks. When I later joined the Uganda Wildlife Authority, it was really the start of my long journey of working with great apes.

One of the early initiatives was the habituation of mountain gorillas and chimpanzees to ecotourism. This work was very important to ensure local communities benefited from the natural resources economically, socially, and environmentally—which, to date, has proved to be the best way of conserving the great apes in Uganda. Ecotourism plays a big role in the conservation of biodiversity, the well-being of local communities, and responsible action on the part of tourists. I conducted several studies that demonstrated the value of these ecotourism efforts. I determined that, early on, ecotourism efforts were often undervalued and that the benefits accrued to tourists were often more than what they were paying. Equally important was some research I conducted that found that when local rural communities recognized the economic value of conservation efforts, they would be willing to protect the forests. This was in part because the financial benefits were better than if they used those lands for traditional agriculture.

A CHANCE CHIMPANZEE ENCOUNTER

When working for the Uganda Wildlife Authority, I was tasked with coming up with new tourism experiences in the national parks, aimed at enhancing tourists' experiences, increasing tourism earnings to go to conservation, and improving the lives of the local communities through a revenue-sharing policy. To prepare for these efforts, I spent a lot of time walking through the national parks and reserves to identify the most compelling features to feature in tourism experiences. On one of the routine walks in Kibale National Park in Western Uganda, my group came across two chimpanzee communities engaged in a boisterous territory fight. We rushed to take cover, but I was able to catch a glimpse of two huge alpha males fighting.

During the fight, one of the two chimpanzees got quite badly hurt and was bleeding profusely from a large gash on his leg. He retreated and I watched as he seemed to start a frantic search among the trees. He stopped suddenly and I observed him pluck off some specific leaves and rub them on his wound. He continued to moisten the leaves with his own saliva and soon the bleeding subsided and eventually stopped. I could hardly believe what I was seeing at the time as I was not familiar with the self-medicinal activities that we now know chimpanzees do. Seeing it happen there in front of me so clearly is what awakened my curiosity about this remarkable species.

CARING FOR CHIMPANZEES

When an opportunity came to join the Chimpanzee Sanctuary and Wildlife Conservation Trust, it was an easy decision. The trust established and manages the Ngamba Island Chimpanzee Sanctuary, which provides refuge for chimpanzees that have been orphaned and rescued from the illegal pet and bushmeat trade. The chimpanzees always come to Ngamba in bad condition: traumatized, malnourished, and infected with different diseases. Despite their past trauma and initial bad physical health, the chimpanzees living at Ngamba have a safe and seminatural environment for recovery, and we have seen them thrive over the course of their long lives. The opportunity to work there was an incredible opportunity to get to know and care for these desperate chimpanzees and to see how resilient they are.

Early on, I performed many different duties at Ngamba. I worked with animal caregivers to prepare meals and clean the chimpanzees' housing. I collected behavioral data on each chimpanzee and helped with their veterinary care as well. But one of the most emotional aspects of the job was to act as a surrogate mother to the orphaned chimpanzee babies who were still in need of a mother's love and care. Along with the other staff, I carefully helped these young chimpanzees grow in confidence as we introduced them to the densely forested areas of the island. More difficult was the challenge of integrating older individuals into some of the existing groups at the sanctuary. Because chimpanzees are naturally territorial, adding new members to a troop is a complex process that requires a lot of care and planning. The process was often dramatic and stressful, but successful introduction of these chimpanzees into new social groups offers them the ultimate chance to live in a community with others of their species, just as they would in the wild.

THE HAPPY ONE

My experiences at Ngamba have blessed me with so many amazing memories of the chimpanzees and their lives. I got to know all of the chimpanzees individually, but sometimes there is a special chimpanzee you remember more clearly than the others. One of the chimpanzees I cared for, and who had a huge influence on me as a person, was Ikuru. While we don't know for sure, the reality is that as a baby, she likely saw her mother killed by hunters. Probably not long after, she was found by a soldier, who tried to keep her as a pet. When Ikuru was about four months old, she was rescued by the Uganda Wildlife Authority, and when she arrived at the Ngamba, she was extremely sick and very distressed. I remember meeting her for the first time. She looked so miserable, rocking herself for comfort. Very early on, I formed a very tight bond with her, and I often accompanied her when we were getting her used to the island's forests that she would eventually be free to roam. I often wonder what that time was like for her. What was she thinking? Did she remember the forests from just months ago when she lost her mother?

Ikuru and I would stop occasionally at different points on our forest walks, and we would groom each other as chimpanzees often do to help maintain social bonds. She had a penchant for undoing shoelaces and cleaning nails. I remember once I had chipped polish on my fingernails and she really did not like that! She carefully took the time to pick the remaining polish off and give me clean fingernails. This became a regular occurrence with not just me but other staff at the sanctuary; Ikuru would "clean" our nail polish and quickly earned the nickname "The Stylist."

Soon it came time to introduce Ikuru to a group of chimpanzees. This step was very important for her social development and for her to learn true chimpanzee behavior. Because of her age and size,

Ikuru

Ikuru was immediately the lowest-ranking chimpanzee in her group and was often bullied by the others. Of course, the staff helped protect her during the integration with the group, but when she was released with others into the forest each morning, she would always return to the indoor bedroom facility in the evening with some slight injuries. I was worried about her and at times I fought the impulse to stop the integration out of sympathy for her. But as her surrogate mother, I knew she would likely benefit from this "tough love" to learn and be strengthened by the process of becoming independent. The integration process was slow and at times we did not know if Ikuru would ever succeed, but after four years of trying, it took just one single act from her to gain all she needed.

The day began for me just like any other day at Ngamba, preparing food for the chimpanzees that would supplement what they foraged in the forest. During feeding times, Ikuru was a frequent target for bullies and food hoarders. She would never defend herself or the food she had been given. We always ensured that she got more food in the evening and morning since we knew she ate very little while in the forest. On this day we threw a piece of fresh jackfruit to Ikuru; she caught it midair and looked around her to see who would try to steal it. Eddie, a high-ranking male, was nearby and immediately came over to her, sensing an opportunity for some extra food. Eddie was much larger than Ikuru and he leaned in menacingly over her. Ikuru froze and looked up at the big male, their eyes locking. A moment passed and then suddenly, slap! Little Ikuru slapped Eddie across the face! The affront caught Eddie completely by surprise; he staggered back and started screaming desperately to the rest of the group. It was such a sight to see the situation reversed by Ikuru, who so often was the victim. The rest of the group all turned at Eddie's pitiful cries, but rather than coming to his aid, they simply stared at the strange situation and at Eddie's dismay, simply returning to their own meals.

From that day forward, Ikuru's ranking in the group started to climb. We saw she had fewer and fewer injuries when she came in for the night. Her confidence grew and she stopped rocking. Her health was regained and today she is a functioning member of the group. She has become what her name has always meant: "The Happy One."

Over the years, I learned many lessons from Ikuru. Her experiences made me understand better that each chimpanzee, just like a human, has a unique personality influenced by both past life history and current environment. This realization inspired me to understand each chimpanzee as an individual with his or her own demeanor. The new experiences of learning about each chimpanzee, through each stage of development and changes in group dynamics, brought a special feeling of discovery to my job and made me more excited about what I was doing each day.

A SEARCH FOR SOLUTIONS

I dedicate a lot of time to thinking about how to prevent more chimpanzees from ending up at the sanctuary. I think it is essential to design strategies that promote harmonious living between humans and wildlife, and much of my work has focused on studying what motivators would get people to protect chimpanzees' natural habitats. I am especially interested in the potential of using private land owned by local residents. Such landscapes are not protected in ways that national forests are, but they may be just as important for conserving the species.

The greatest threats facing chimpanzees today are the illegal bushmeat trade and rampant habitat loss. There is currently an unsustainable and thriving demand for chimpanzee meat in markets across Africa, contributing to the overall demand for bushmeat, which is now a billion-dollar industry. Coupled with the destruction of key chimpanzee habitats, due to the expansion of agricultural and other development activities, this demand puts further pressure on already stretched landscapes. Humans and chimpanzees are forced to live closer together than ever, which can result in retaliatory killings if chimpanzees hunt or graze on crops and farmland. Shrinking habitats and competition for space also mean humans are exposed more regularly to wild chimpanzees, increasing the interest and demand as a resource for either food or money. Closer proximity to each other also increases the risk of disease transfer, as chimpanzees are extremely susceptible to many human diseases.

As huge as these challenges are, I believe it is possible to overcome them with a systematic approach focused mainly on education. It is possible to reverse the tide and change the culture of bushmeat consumption by having people consciously decide to manage their own lifestyles. By addressing related issues, such as decreasing poverty with improved agricultural practices, the solutions to manage these threats become even clearer. There is a great need to start working on these problems, not only at an individual level but also at a landscape level. There is so much to work on, but we should not ignore the past and the successes and failures that we and others have experienced. I am proud of my past achievements and hope they can continue to grow in ways that will help chimpanzees and communities alike because they are examples of what people can do when they set their minds to it. These include helping to build the Ngamba Island Chimpanzee Sanctuary as a model organization, promoting innovative and effective conservation education approaches, and promoting the training of women in conservation.

One of the things I am most proud of is my work to influence policy that affects thousands of apes around Uganda, such as my work on the National Environmental Act of Uganda. My hope is that these protections for wildlife continue to strengthen and help to enhance community conservation efforts. But there is still a lot of work to be done, and more of that has to do with sustainability. It is hard to talk about rehabilitating the chimps without thinking of sustainability. A lot of support is needed from both the government and private sectors to not just protect the wins gained so far but also to help realize more impactful results with this kind of work. Future progress remains a collective responsibility.

FINDING HOPE AND COMPASSION FOR THE FUTURE

Working with chimpanzees has had a tremendous impact on my life. The success with the welfare and management of the chimpanzees at Ngamba inspired me to think in the long term. I have been inspired by the chimpanzees to do more with my life in order to serve this planet. I discovered that I have a particular purpose in this world, so I try to do more in order to make the world a better place. My hope is that this work will contribute to the good others are doing and minimize the destruction that would occur if I do not do my part.

I would like to see more and more people aspire to become primatologists since the demand for them continues to grow. For anyone wishing to become one, I have just a few things to say. Always remember how you would want to be treated, as it is the same thing animals in their care would wish for: compassion and understanding. Hard work and being passionate about what you do will set you apart from the rest. And not forgetting to be creative and innovative when it comes to dealing with the different situations this job will put you in will also enhance your work and inspire others.

I started by talking about how it feels every time I arrive at Ngamba and so I will end here by talking about how I feel each time

I leave. To me it is always a cocktail of emotions. On one hand I am happy and content from the experiences of meeting and catching up with friends and remembering special moments we have shared. But on the other I cannot help but feel sad when it's time to depart because no matter how long the visit has been, it always feels too short when it is time to go. It reminds me of how short life is and how we must strive to make the best of the time we have. I always wonder when I will be back to the sanctuary again, and if my friends will be there waiting, though they always live in my heart. That's the kind of feeling I want everyone to leave with, knowing that you were part of the moments, both large and small, and you are able to walk away in fulfillment.

3

RICHARD WRANGHAM

Dr. Richard Wrangham is an anthropologist whose work has primarily focused on chimpanzee ecology, nutrition, and social behavior. In 1987, he founded the Kibale Chimpanzee Project in Kibale National Park in Uganda. There, a group of fifty chimpanzees in the Kanyawara region of the park is studied on a daily basis. With his wife, Dr. Elizabeth Ross, he founded the Kasiisi Project, which supports conservation and education activities, working with communities that neighbor the park. He is currently a research professor of biological anthropology in the Department of Human Evolutionary Biology at Harvard University.

FROM ORNITHOLOGY TO PRIMATOLOGY

A sense of inspiration came to me the first day that I saw chimpanzees in the wild. But I get no primatological credit for reaching that point. As a boy growing up in an English village, I had good long-distance eyesight, a father who liked bird-watching, and an introvert's desire to be alone. Britain was full of bird-watchers, so I became a bird-watcher. I was less committed than many, but I was enthusiastic. So it was birds that took me into the wild. During two adolescent summers I spent a few weeks on Fair Isle, one of the remotest and most beautiful islands in Britain, learning how to identify, ring, and study birds under the guidance of Roy Dennis, who went on to lead conservation efforts in Scotland and elsewhere. Birds also introduced me to science, and a note I wrote about a bittern that I found outside its normal habitat during an exceptionally hard winter became my first publication.

At Oxford University I studied zoology but, during my three years there, I only attended a single lecture on primates. The Nobel Prize–winning biologist and ornithologist Niko Tinbergen was there, mostly supervising graduate students. I talked to him about a PhD and he suggested I study cattle at his site in northern England. Cattle were nice, but I preferred something more directly related to human behavior. I thought of humans as diurnal, social carnivores. Banded mongooses were also diurnal, social carnivores and so I hoped to study them in Uganda. It did not work out when I could not raise the necessary funds. I asked my tutor, Harold Pusey, for advice. He was an expert on the chondrocranium of frogs. "Well," he said, "my daughter Anne is going to work for Jane Goodall studying chimpanzees in Tanzania. Why don't you write to Goodall and ask if she would like to have you, too?" I knew who Anne was. She was one of about twenty-five fellow students in my zoology class. I did not know who Jane Goodall was.

Jane was developing a research team at Gombe National Park, and she had a grant to help her. She could not have been more welcoming. In November 1970, six months after graduating, I arrived in Tanzania. Chimpanzees were in the camp on my first day there. Jane assigned me to collect data on sibling relationships. It was an excellent time to be doing it. Until 1969, Jane and her assistants had spent relatively little time following the chimpanzees outside the provisioning area, where bananas were provided several days a week and where close-up observations could be made in unparalleled detail. But in 1970 the banana-feeding was being reduced (it is not done today) and researchers were spending more time following chimpanzees wherever they went. There were four or five of us there for most of the year, and when we met in the evening the sharing of the day's experience could be exhilarating. Much was new.

The privilege of being in Gombe was huge. I had no obligations other than to simply spend time watching these individuals and record what I saw. Spending so much time with my seven sibling "targets" allowed me to think about individual strategies from a very personal perspective. Connections with environmental pressures seemed clear. As the year rolled by, different species of fruit came and went. More food, and bigger food sources, led to bigger parties of chimpanzees. Less food, and suddenly the chimpanzees were spending a lot of time in very small subgroups or even alone. Surely these changes would have a lot of meaning for the chimpanzees?

Gombe seemed like a gold mine of opportunity. I was very lucky that during my final months there, Robert Hinde and David Hamburg visited and decided to support me. I became Hinde's PhD student at the University of Cambridge, supported by a grant that Hamburg had obtained. Years later, when Jane stopped directing the research at Gombe, Anne Pusey took over to lead the science. There was inspiration in the air in 1970, the year we both joined the team. We all felt it.

THE PLEASURE, THE SATISFACTION, AND THE THRILL

Ultimately, my fascination with chimpanzees comes out of my concern for our own species. They help us understand ourselves. But the proximate reasons are no less important. Three are prominent.

One is the sheer aesthetic pleasure of observational fieldwork. You feel it before dawn at the start of a day to be spent following a chimpanzee. Relaxing after arriving at the nest-site, lulled by the

night sounds, waiting for the first stirrings in the nest, the antici-
pation during that calm time can be delicious. Will we have a day
vibrant with forest life? A family of giant forest hogs. A black-and-
white casqued hornbill mother poking from her mudded hole.
And might the soap opera of chimpanzee social relationships take
another turn? Will we get to see something truly new, strengthen-
ing an insight, raising a new idea, showing a new capacity? There is
always a sense of possibility, the chance that momentarily we will
come across something exciting or beautiful or absorbing. Or will
we have a long day slogging so clumsily through mud, swamps, hills,
and rain that we are mostly too far behind our target chimpanzee
to catch the interesting details? Even then, a day with chimpanzees feels special. The privilege of being admitted for a few intense
hours into a once secret world is very real. The reward might be no
more than an eventual half-hour spent with a mother and infant
son relaxing on their own, idly playing on a bed of leaves dappled
by the late afternoon sun. The two quietly enjoy each other's
company—he rehearses little routines, she gently tolerates his
repeated bouncing on her affectionate body. The sight brings a stab
of familiar emotions and reminds us that life goes on like this without us, echoing a scene that has been played out every day for mil-
lions of years, including by many unknown species, and by some
known to us only by scraps of fossil bone.

The second reason I like studying wild chimpanzees is the
satisfaction of helping. These apes are fellow beings with minds
sufficiently similar to ours that we can feel exceptionally sympa-
thetic to their difficulties. For us in Kanyawara our most direct aid
for chimpanzees in our community occurs when we intervene to
stop the suffering caused by a snare. Seeing a particular individual healed who would otherwise be wounded or dead is rewarding
in a very personal way. But the opportunities for a snare intervention are rare, thank goodness (though at about one per year, they

are still frustratingly high). Mostly, the goal of helping is more remote because it comes from conservation efforts. The fact that research tends to promote long-term conservation of habitats and populations brings its own powerful motivation for maintaining a field study.

Third is the intellectual thrill. Time in the forest spent watching chimpanzees brings ideas about the constraints and opportunities on their behavior. However, ideas alone are not worth much. Their value comes from being tested. The daily amassing of data by our research team is not an end in itself, but it almost feels like that because it is so exciting to know that with it new knowledge will be gained.

The pleasures of spending time in nature, the ability to contribute to welfare and conservation, and the satisfaction of contributing to our knowledge of a unique species make up a phenomenal triad. On their own they would be enough to motivate a lifetime spent in studying chimpanzees, just as they keep people entranced in research on other charismatic species. Some of us working with apes, however, feel an additional draw. Our studies help understand our own species. That was why Louis Leakey invited Jane Goodall, Dian Fossey, and Biruté Galdikas into the field. In the almost sixty years since then, his intuition that ape research would benefit human self-knowledge has been thoroughly justified.

Leakey could not have guessed that chimpanzees would prove to be more closely related to humans than gorillas are. He died in 1972, more than a decade before the publication of the DNA hybridization research that led to the idea being taken seriously. But Leakey learned of some of the stunningly unexpected similarities in behavior between chimpanzees and humans, including tool-making, hunting, food-sharing, and bonding among males. I was fortunate to meet him in 1971. He seemed to be thinking, as we know today, that the similarities were likely features that chimpanzee and human lineages had shared in parallel since our common ancestry. He was thrilled about it. Had he known the evidence that makes chimpanzees even more directly relevant than bonobos or gorillas in reconstructing the human past, he would surely have been even more pleased.

KAKAMA: THE CHIMPANZEE WHO LOVED A LOG

Chimpanzees in captivity regularly demonstrate abilities that raise questions about the nature of the species. Reports of such behaviors as altruism, cooperation, partner selection, and postconflict mediation can be intriguing because of their rarity in the wild. What about Viki, the chimpanzee who grew up living with Cathy and Keith Hayes in the 1950s? Viki would play not only with a real pull toy but even with an imaginary one. Cathy described how Viki would hold her hand behind her as she walked, making the sound of the pull toy trundling along at the end of a string. Quite remarkably, Viki changed the sound when the pull toy moved from being on a wooden floor to a carpet. She would make the change not when she herself reached the carpet but when the imaginary toy, just behind her, did. The wild offers few opportunities for such elaborate behavior. So it is hard to know whether captive chimpanzees have acquired their human-like abilities as a result of their atypical living conditions or whether, on the other hand, wild chimpanzees are equally mentally inventive as captive chimpanzees but simply are more constrained by the exigencies of their lives. Imaginative toy-play in the wild suggests the latter.

Kakama was born in July 1985 and was the oldest chimpanzee in the Kanyawara community whose age we knew for sure. Gilbert Isabirye-Basuta had been studying the chimpanzees since 1983 and knew Kakama's mother, Kabarole, well. She was easily recognized

Kakama

by her pointy arm, being one of two mothers who had lost her hand to a snare. Kabarole became habituated early. By 1993, I could follow her when she was alone with Kakama.

By then Kakama was almost eight years old and he had no younger siblings to play with. But now Kabarole was pregnant at last. She had always walked slowly, perhaps because of her missing hand. She tended to abandon the big groups that would travel long distances to reach rich food patches. Her pregnancy seemed to make her walk even more slowly than before. She and Kakama spent more time alone together than ever. Although he was fast approaching adolescence and was at an age when some young males start accompanying older males, Kakama stayed with his mother. The relationship between the two was exceptionally close. Sometimes Kakama would sprawl over Kabarole's body and suck Kabarole's lower lip for minutes at a time—up to eight minutes on one occasion. The behavior suggested a deep need to be loved. I have not seen it in any other wild chimpanzee families, but it occurs in captivity among juveniles growing up together without their mothers.

One day during Kabarole's pregnancy, she and Kakama were resting on the ground while I sat a few yards away and watched them. Kabarole got up and ambled off in the direction of a fruiting tree five minutes' walk away. Kakama turned and started to follow the direction taken by his mother. He bounced along for a short distance then flopped onto a rotting log about the size and shape of a full hot-water-bottle. Putting his hands around the log, he kept it close to his body and instead of getting up to walk, he somersaulted forward a couple of times. He ended up standing on all fours holding the log to his belly. Kakama walked onward, taking the log with him. Mostly he kept it in place on his lower back by curling his hand behind him. Once he held it by one end in his right hand, walking tripedally, while the log bumped up and down on the rough ground.

He caught up with Kabarole and followed her downslope together, sliding through the scrub too fast for me to keep up with them.

I found them again later from hearing the rustling of branches. They had climbed into a tree. By the time I had worked my way to a place where I could see them, both were already fifteen or twenty meters up, sitting and chewing the cherry-sized fruits. When Kakama climbed to a new feeding position, however, he revealed that he still had his log with him. He would place it carefully on the branch next to his body and then make sure to pick it up and take it with him whenever he moved. I had not seen such a behavior before. Chimpanzees sometimes kept the skin of a prey animal they had caught with them for a few hours or even overnight, and in Gombe they had sometimes stolen clothes or towels from the researchers and kept them a long time, also for chewing on. However, there seemed to be nothing special about the log that would explain why Kakama was keeping it with him. I wondered how much he cared about it.

The question was soon answered when the log fell from one of its precarious perches. Kakama climbed down, retrieved the log, and returned to his place in the tree crown. He resumed feeding, still accompanied by his strange possession. Eventually Kabarole stopped eating. As she often did following a satisfactory meal, she spent a minute making a nest and then lay down to rest. Not long afterward Kakama did the same. Looking from below, I could not see into his nest. I could occasionally see what he was doing, however, when the log appeared above him, supported by his feet. Kakama was playing the airplane game with his log, much as chimpanzee and bonobo mothers and human parents sometimes do, lying on their backs and holding their offspring above their bodies. After fifteen minutes, and apparently accepting that they would likely be staying in the tree for a further rest period, Kakama made a smaller tree nest than I have ever seen a chimpanzee make, certainly too small to accommodate himself. It was the right size for his log, however. He lay the log in the nest. He then left the log and rested in his own nest an arm's length away.

After our research team was able to observe the Kanyawara families closely we learned that juveniles have a cultural practice of carrying objects. Mostly they carry sticks, which they often take into their nests during daytime rest periods. Sometimes their chosen objects are rocks. They rarely carry logs, however, and I have never heard of another wild chimpanzee making a nest for one of its toys.

So why did Kakama make a small nest and put his log in it? He seems to have been treating his log like a doll. Perhaps his eight years without a sibling made him long for companionship. Possibly he realized that Kabarole was pregnant and that he would soon have an infant sibling. We may not be able to answer those questions, but Kakama's behavior at least suggests that the kind of imagination shown by Viki is likely to be present in the wild—even if there are few opportunities to show it.

LIVING LINKS

Generating ideas about where we come from is a big deal. Humans are the only species capable of asking why we exist, and we happen to be living during a period when we are generating newly confident answers at astonishing speed, at least compared to the probable hundreds of thousands of years since humans have been able to share their thoughts on such matters. Charles Darwin gave us the framework. Now evolutionary anthropology is producing the details. Within a few decades we will surely have a story of where we came from, and why, that will last for all time.

Ape studies contribute not only to those lofty goals but also to a more immediate understanding of ourselves as biological entities.

Many people have been reluctant to see human behavior as influenced strongly by our biology, but the apes force a more realistic view. The similarities superbly show chimpanzees and other apes as living links between humans and other species, a vivid reminder of our evolutionary background. They remind us that to claim that human social universals are merely a product of culture, as is still often done, represents a cavalier attitude to the possibilities and even the likelihoods. During the coming decades we will learn enough about the mechanisms underlying social behavior in apes and humans, including neural and endocrine systems, to assess how similar apes and humans truly are. Those advances will give us a much sharper understanding of why some aspects of our social behavior are apelike and others not. Meanwhile, the patient assessment from field studies of the adaptive significance, individual development, and causal control of ape social behavior gives us the core information that will contribute to a new understanding of the biological nature not only of the apes but of humans, too.

In 1993, the writer Dale Peterson kindly invited me to a New Year's Day party. Over champagne I told him about the emerging information on contrasts between chimpanzees and bonobos in the levels of violence they displayed. I shared some ideas about how the species' differences might be explained and what the explanations might mean for understanding the evolution of aggressive behavior in humans. He suggested we write a book on the topic. I was leaving a few days later to spend six months in Kibale. Between fieldwork, teaching, and raising a family I never thought I would have time to write a book on my own. But Dale was an established author who knew a lot about primates. The idea of collaborating with him was wonderful. We shook hands on it. Three years later, *Demonic Males* was published. In my view, our book took a fairly orthodox approach to understanding a particular type of social behavior, namely male aggression. While it was designed for a general readership, the aim was to be accurate and clear with respect to the science. Dale added the color. If I describe how and why I was surprised by some of the reactions to this volume, I suspect I will be setting myself up to receive more of what took me aback originally. I will be less startled now, however.

I recognized that the title was provocative. It was not my first choice, but publication is teamwork and I accepted the idea cheerfully enough. The merit of the phrase *Demonic Males* was that it conveyed one of the book's core themes: the high rates of violence that humans sometimes show, especially in wars, make us an unusual mammal. One of our goals was to explain why humans are a member of that rare set of species in which adult males collaborate to deliberately kill members of neighboring groups. Had I been educated as an anthropologist I would have had a deep knowledge of the discipline's intellectual and social rifts, but even with my background as a biologist and primatologist I knew of strong disagreements with regard to evolutionary thinking. Like Jane Goodall before me, I was nevertheless surprised by the hostility of responses from people who disputed the view that chimpanzee behavior has anything to teach us about the evolution of human violence.

The critiques were varied. Chimpanzees were claimed to be more peaceful than Dale and I had described. Alternatively, even if some chimpanzees were admitted to be occasionally violent, those bad apples were said to be only a meaningless subset because they had supposedly been induced into the habit of killing by having their natural lives disturbed by human activity. Another idea was that even if chimpanzees are sometimes aggressive, in *Demonic Males* we had exaggerated the scale and intensity of human violence, so the comparison between chimpanzees and humans was wrong. Others dismissed the idea that chimpanzee violence offers useful opportunities for comparison with humans given that bonobos are less violent and equally closely related to us. A different kind of critique came from those who accused us of treating "male demonism" as the defining characteristic of our species.

These kinds of questions and criticisms can be made in various styles. I was prepared for constructive disagreement, just as I still enjoy talking about such issues with people who debate in an open-minded spirit of enquiry. By contrast, I was not prepared for the readiness with which some used *Demonic Males* to treat me as a member of the wrong faction. These critics regarded the book as evil. Contrary theoretical perspectives are legitimate, but no favors are done to science when ideas that challenge are treated as the enemy. I would like to be able to say that polarizing responses to scientific ideas are fading from the intellectual scene. Our species being what it is, however, that may be optimistic. We can still hope.

4

JOHN MITANI

Dr. John Mitani is a behavioral ecologist and professor emeritus in the Department of Anthropology at the University of Michigan. For several years, he has studied the behavior of an exceptionally large community of chimpanzees at Ngogo in Kibale National Park in Uganda. Kibale is a mid-altitude rainforest and home to seven species of diurnal primates. The Ngogo chimpanzees have been subjects of several documentaries, including Rise of the Warrior Apes and Disneynature's Chimpanzee.

A SERENDIPITOUS START

I have studied wild chimpanzees for over thirty years and for most of those, I have studied a huge community of chimpanzees at Ngogo in Kibale National Park. My route to Ngogo, and the chimpanzees there, was long, circuitous, and full of lucky breaks, so I am a strong believer in serendipity in science.

My interest in animal behavior began in high school when my parents bought a National Geographic book for my younger brother called The Marvels of Animal Behavior. It was full of spectacular photographs, with articles written by leaders in the field: Dick Alexander, Jack Bradbury, John Eisenberg, Steve Emlen, Hans Kruuk, Gordon

Orians, and George Schaller. I was particularly drawn to a paper on baboons by Irv DeVore. I stole the book and left home to attend the University of California, Berkeley. Sherry Washburn, DeVore's mentor and one of the founders of primatology, taught at Berkeley, and I was fortunate to take classes from him. I was hooked.

I still had baboons on the brain when I started graduate school at the University of California, Davis, where I was supervised by Peter Rodman, one of DeVore's students. Peter had just completed one of the first successful studies of wild orangutans, and he convinced me to study Bornean primates. After a few failed projects, I completed my PhD investigating the spacing and mating behavior of the two Asian apes, gibbons and orangutans. I continued that line of research for twelve years, incorporating work on their vocal behavior during a postdoctoral research position with Peter Marler at Rockefeller University.

Peter Marler is why I study wild chimpanzees today. He spent most of his career unraveling how male birds learn to sing their songs, but he also conducted fieldwork in East Africa on the vocal behavior of primates. This included work with chimpanzees at Gombe National Park with Jane Goodall. Remembering his time with the Gombe chimpanzees and the interesting things they had to say to each other, Peter persuaded me to start work on the vocal behavior of chimpanzees. When deciding where to go, I recalled meeting Toshisada Nishida in Japan. I asked Toshi whether I could study chimpanzees at the Mahale Mountains National Park in Tanzania, where he had been working for many years. He agreed, and I began my career as a chimpanzee field researcher.

I worked at Mahale for five years. I relished my time there and learned a great deal from Toshi, who was a keen observer of chimpanzees and shared his wealth of knowledge with me. I wouldn't know half of the things that I know about chimpanzees if it were not for Toshi. He passed away in 2011, and I still miss him and his sage counsel and friendship. Under Toshi's tutelage, I became a committed chimpanzee fieldworker. Like him, I came to savor my time with chimpanzees.

Another serendipitous event led me to the Ngogo chimpanzees in Uganda. In 1994, I was at a conference for researchers studying great apes with David Watts. David was planning to conduct a pilot study of the Ngogo chimpanzees the following summer, and I asked if I could tag along. I wanted to see Kibale National Park, as it was a world-renowned primate field site started by Tom Struhsaker, who I knew. It took only a few weeks at Ngogo to realize that the site was unusual. There were many chimpanzees, and they were everywhere! At the end of my two-month stay, I asked David whether I could join him to start a project at Ngogo. He agreed, and the rest is history. It's been an enormously productive collaboration.

The dirty little secret about chimpanzee fieldwork is that it can be boring, as chimpanzees rest and sleep a lot. But if you are patient and commit the time, they will always surprise you by doing something novel and interesting. This is what Nishida meant when he wrote, "Chimpanzees are always new to me." It's why I continue to find the study of chimpanzees so rewarding. Intriguing new observations continually lurk around the corner even after thirty years of study.

HARE'S LESSONS

I have studied chimpanzees in two communities: the M group at Mahale and the Ngogo community in Kibale. There have been many memorable chimpanzees at both sites, but if pressed to choose, one chimpanzee has had a particular impact on me. His name is Hare, and he has been a key player in the stories that we have told at Ngogo. When we started our studies there more than two decades

Hare

ago, Hare was already an adult male about twenty-five years old. He was unusual physically and easily identifiable because of his cleft palate.

Hare's distinctive appearance made him easily recognizable from the start, and he made a significant impression on us not too long after we began our work. As is typical at other field sites, the Ngogo chimpanzees fled from us upon first contact. It was a chore to get them used to us, and we spent a long time habituating them to our presence. Hare was the very first chimpanzee to tolerate mine.

During the second year of study at Ngogo, I watched as the chimpanzees fed in a huge fig tree. I was waiting below trying to identify individuals in the tree. After several hours gorging, they descended. As Hare came down, I got behind him and started to follow. I expected him to run away as he had been doing the past year. Instead, he walked off slowly—ten meters, twenty, then fifty. I continued to follow behind thinking that he would take off at any moment. He didn't. He continued to amble on through the forest for a kilometer until he arrived at another fruit tree. When I realized what was happening, I almost started to cry. After working so hard for so long, it was tremendously gratifying to have gained Hare's trust. Once he fell in line, other males, one by one, started to habituate to my presence over the next several months. We were off to a good start to our study thanks to Hare's trust in me.

Over the years, Hare has informed us about male chimpanzee behavior in many ways. He was among the most skillful hunters, but what was more interesting than the hunts themselves was what Hare did with the meat he obtained. He, along with other males, shared meat voluntarily with others. He didn't dole out his bounty randomly but instead gave meat selectively to others, mostly to his friends and allies. Hare reinforced his friendships not only by sharing meat but also by forming coalitions with his partners. Through studies of Hare and other male chimpanzees, we have also documented reciprocity in meat-sharing and coalitionary behavior.

During his prime, Hare was very high ranking and played the male chimpanzee social game just like the others did. But he was unusual in some important ways. Male chimpanzees are very aggressive, and as they strive to obtain high dominance status, they will use any and all opportunities to attack lower-ranking individuals. Hare, however, had a softer and gentler side not often displayed by male chimpanzees.

Several years ago, David Watts witnessed male chimpanzees at Ngogo savagely attack one of their own: Grappelli. After the attack, Grappelli crawled into a tree. The other males continued to threaten and lunge at him. As they did so, Hare moved to a spot between the seriously wounded Grappelli and his would-be attackers. He seemed to be trying to protect Grappelli. I witnessed a similar situation a few years later when Halle, a young adult female, had given birth for the first time. A large group of male chimpanzees including Hare encountered Halle and her new infant. The males beat her relentlessly, but Hare did not participate. After the brutal attack, Halle, like Grappelli, managed to scramble into a tree. Again, Hare put himself between her and the other males to fend off further attack.

Hare's kind and calm demeanor was manifest in another way. Some male chimpanzees show an avuncular quality by spending time, grooming, and playing with youngsters. Most males display these behaviors late in life when they enter what I call "retirement." But some males just seem to be this way given their temperament and personality, irrespective of age. Hare was one of those chimpanzees.

Hare was also remarkable in terms of his response to the loss of a close ally. Early on it became clear that Hare had a good friend, Ellington. When Ellington disappeared in 2002, Hare's behavior and demeanor changed dramatically. Prior to this, Hare was a very gre-garious chimpanzee, who was always around if there was a big party of chimpanzees moving from one spot to another. He would always be in the mix and a central part of the group, but this changed after Ellington's death. Hare became a recluse and did not seem to want to be around other chimpanzees anymore. When he did appear, he wouldn't stay long. Hare appeared to mourn the death of his good friend Ellington.

Unfortunately, Hare died in 2017. Human observers who knew Hare mourned his passing as he did Ellington's. He lived a good and long life, though, and he leaves a legacy. Hare fathered five infants, including two males, who continue to reside at Ngogo. I will watch Hare's sons with continued interest and be fascinated to learn their fates. I will never forget their father and the many lessons he taught me about chimpanzee life.

WHY I STUDY CHIMPANZEES

There are three things that motivate me to study chimpanzees in the wild today. They all relate to fieldwork and can be characterized as methodological, empirical, and spiritual.

Methodologically, I am committed to fieldwork. I believe quite strongly that there is no better way to learn about chimpanzees than by spending long stretches of time with them, patiently watching and recording what they do. The work can be hard and tedious, but the rewards make it worthwhile.

I also want to contribute to our empirical understanding of chim-panzee behavior. Chimpanzees are fascinating, but coming to grips with what they do is difficult. Understanding their behavior is like putting together a giant jigsaw puzzle whose pieces do not always fit. It takes time and effort. Occasionally, one has an aha moment when the bits and pieces of the puzzle come together. Admittedly, these moments are few and far between. But when they happen, there is nothing more satisfying. I have had a few of these moments at Ngogo, and they furnish the incentive to put in long hours with the chimpanzees there.

My first aha moment at Ngogo came the very first summer I visited. It did not take long to realize that there were chimpanzees everywhere. A few years later, it dawned on me that the male chimpanzees at Ngogo formed subgroups and lived in "neighborhoods" in the same way that female chimpanzees do in East Africa. Another light bulb went off shortly thereafter as we continued to observe male chimpanzees hunt red colobus monkeys. As I described for Hare, after capturing prey, instead of sharing meat randomly with others, the chimpanzees did so selectively with their friends and allies. Years later, after they had decimated the local population of red colobus monkeys, the chimpanzees switched prey, hunting other vertebrates in earnest that they rarely touched before.

A startling revelation came in the summer of 2009 when the Ngogo chimpanzees began to inhabit a new area far to the northeast of their territory. In the past, they had only used that area during territorial boundary patrols. The area was otherwise a forbidden zone. During that summer, though, they showed no reluctance to travel there. Males, females, and youngsters frequented the area, announcing their presence loudly by calling out frequently. They acted as if the place was their own. This was head scratching at first. But after a few weeks, when I was walking back to camp one day, I realized that the Ngogo chimpanzees had killed many neighbors in the area. By doing so they had reduced the coalitionary strength of the neighboring group to such an extent that they had then made a land grab, usurping the territory that once belonged to others.

Finally, a spiritual reason motivates me to study chimpanzees today. It is deeply personal and difficult to explain. The best way to understand is to imagine the following situation. I am out alone with a large party of chimpanzees. There is an abundance of food, and there are perhaps forty, fifty, maybe even sixty chimpanzees together. And it's a glorious, dry, sunlit day. The chimpanzees have settled down to rest and socialize. Everywhere I look, there are chimpanzees on the ground. A few adult males groom. Mothers relax and begin to doze off as their kids start to play. Some wrestle, laughing noisily in the process. Others chase each other around small saplings and then drop to the ground with a thud. I have experienced countless scenes like this, and as I survey it, I am overcome with utter joy. I am astonished that the chimpanzees permit me to be a part of their world, and I feel that I am the luckiest person on earth. Moments like these may be the most important reason I continue to study chimpanzees.

COUNTING CHIMPANZEES TO CONSERVE THEM

As described by many of the contributors to this volume, chimpanzees face many threats to their survival. Habitat destruction due to humans, recurrent outbreaks of infectious disease, and a thriving bushmeat trade continue to decimate populations of wild chimpanzees. Solutions to these complex problems are difficult and elusive, leaving wild chimpanzees in a precarious state. Protecting chimps in the wild is made even more difficult because of another problem that has, for the most part, been overlooked. Designing effective management plans to conserve chimpanzees requires that we know where they live and how many there are. All too often, however, we lack these fundamental data.

Kibale is home to several primate species living together at very high densities. Primate field research there was started fifty years ago in 1971 by Tom Struhsaker. Over the years, many researchers have flocked there to conduct research on the ecology, behavior, and conservation of primates. With so many people running around the area for so long, one would think that we know how many chimpanzees live in the park. Nevertheless, we have had no idea until recently.

The reason for this is easy to understand. Chimpanzees live in very low densities and range over extremely large areas. Several years ago at a professional meeting, I was dumbfounded by a colleague who came up to me and pronounced with considerable certainty, "There are no chimpanzees at Ngogo and in the surrounding area." He had come to this conclusion based on some hasty counts of chimpanzee nests in the area. My encounter with this colleague occurred after I had been studying the Ngogo chimpanzees for over ten years and had written about them, noting multiple times the extraordinarily large size of the community. I was baffled and at a loss for words. Reliable information regarding chimpanzee numbers typically demands long-term research on specific communities and populations.

Happily, the question of how many chimpanzees reside in Kibale is now being determined through some innovative and hard work by my colleague Kevin Langergraber. Using a novel and noninvasive genetic capture-recapture methodology that relies on the collection and study of fecal samples, Kevin has shown that there are more chimpanzees in Kibale than anyone would ever have imagined. Some of them are likely to live in communities that exceed one hundred individuals. This is an astonishing finding. Until now, no one has had a hint that there were so many chimpanzees in Kibale. Yet without these data, it will be impossible to design and implement effective conservation strategies. We have a lot of work to do. It will require time, effort, and boots on the ground.

CREATING CONNECTIONS

It is easy to surround yourself with people who share your interests and convictions, but to effect real change for chimpanzees, we have to reach out beyond our own community and engage with people outside of our field. We rely on public and private funding sources to support our research, and those who hold the purse strings need to know what we have learned and why it is important. Moreover, our efforts to protect chimpanzees in the wild depend on disseminating our findings widely and effectively. I have always felt that it's important for others beyond the academy to know what I do and why I do it.

A few years after initiating our study, filmmakers began to approach us about filming the Ngogo chimpanzees for nature documentaries. At first, David Watts and I were reluctant to have anyone come, as it would detract from our work. After repeated requests, however, we finally relented. A key moment occurred when Alastair Fothergill, the former head of the BBC's Natural History Unit approached us about filming the Ngogo chimpanzees for his *Planet Earth* series. We agreed, in part, because it was a serious effort. Alastair was willing to commit at least two months of filming at Ngogo.

A few years later, Alastair was ready for a new challenge. Based on the success of *March of the Penguins*, Disney formed a new studio, Disneynature, which was commissioned to make nature films and release them in movie theaters. Alastair made the first film about chimpanzees. He was pleased with the footage he got at Ngogo for *Planet Earth* and approached us about filming the chimpanzees there again for the movie. When he came to Michigan to talk to me, he said that he wanted to make a motion picture that showed chimpanzees as they truly are. Having no reason to doubt him, I agreed, perhaps naively. What ensued was a four-year project from start to finish, and the film, *Chimpanzee*, was released in 2012.

A few years after that James Reed started to call me. He had worked on *Chimpanzee* and had become intrigued by the Ngogo chimpanzees. He wanted to make a film about them. After several months of phone calls, many of which I answered rudely, I finally relented. In retrospect, I should have been as cooperative as a male chimpanzee at Ngogo. James is a master storyteller and put together a brilliant film about the Ngogo chimpanzees, *Rise of the Warrior Apes*. His efforts received critical acclaim, and the film has won prestigious awards. In sum, I have tried my best to work with the production crews on these film projects because I believe these films represent one the best ways to engage the broader public and inform those in the lay community about chimpanzees and their behavior.

REFLECTIONS AND RECOMMENDATIONS

What lessons are to be learned from all of this? I have been incredibly lucky. I have been to places and seen and done things that most people can only dream about. I have also been lucky to have had a wonderful group of mentors, colleagues, and friends, all of whom have furnished sound advice. So, at the end of the day, lean on those around you and take advantage of the opportunities you are given. Along the way, if you are lucky like me, you may learn a few things.

Ours is at heart an observational science. While major advances are being made in the study of animal behavior through the development and adoption of new technology, there is no substitute for direct observation. Many of the major discoveries about the behavior of chimpanzees and other animals have been made only after years of patient, time-consuming observations in the wild. Occasionally, opportunistic observations, instead of carefully executed scientific studies, have resulted in profound insights into chimpanzee behavior. Think of Jane Goodall's observation of chimpanzees in the wild making and using tools and how this changed our view of humans as unique creatures in the animal world.

In addition to spending time with your animals and watching them closely, another bit of advice I will offer is something that I preach, but as my wife and students can attest, do not always practice: be patient. In a day and age when we are rewarded for quick fixes, solutions, and answers to problems and questions, the study of chimpanzees and other primates demands patience because these long-lived animals give up the secrets of their lives to us as human observers only very slowly.

This is a very small and esoteric field of study. We are extraordinarily lucky to be able to do what we do. So cherish the chances you have to study your animals. And do what you can to ensure that others have similar opportunities in the future. One way you can do so is by devoting time and effort to the discipline itself and helping others who belong to our community. Give back any way that you can. The study of animal behavior will thrive only through continued participation by a new cohort of dedicated researchers. Encourage students who are starting out. You will be surprised how a simple, nonrandom act of kindness can go a long way to help a budding colleague.

5

CAROLINE ASIIMWE

Dr. Caroline Asiimwe heads the veterinary department at Budongo Conservation Field Station, which is recognized as the national chimpanzee health monitoring center for Uganda. Nestled alongside Lake Albert, along the border with the Democratic Republic of the Congo, Budongo Forest is home to an amazing array of flora and fauna, including approximately eight hundred chimpanzees, many of whom have been under study since the 1960s. She leads a group of conservation rangers, many of whom are former poachers, to protect the densely forested ecosystem and to mitigate the human-wildlife conflict in the area.

MOTIVATED BY EMPATHY

As a veterinarian and wildlife conservationist, my focus is almost always on the animals but throughout my life and my career, I have learned the importance of understanding the relationships between people and wildlife. Only by collaborating and exchanging information, between academics and forest-edge communities with their indigenous knowledge, can we best hope to protect local ecosystems and serve local people. It's an idea that has origins very early in my life.

I was only eight years old when I first encountered chimpanzees in Uganda. But they were not among the wild chimpanzee groups that live in the many rainforests in this country; I was visiting the Entebbe Zoo. Now a much different place, and renamed the Uganda Wildlife Conservation Education Centre, back then the chimpanzees and other animals were in small, barren cages. Visitors would hit the bars of the cages over and over again, and the chimpanzees would respond by running around and making all sorts of loud noises. I must admit that we all thought this was the way that chimpanzees entertained us visitors. I recall that I stood looking at the chimpanzees and watching visitors throw scraps of food and pieces of trash to them. One man threw a cigarette at the chimpanzees and the big male quickly retrieved it. I watched as he carefully opened up the cigarette and started to eat the bitter tobacco contents. At that moment I felt terrible for the conditions that the chimpanzees were in. Was it because I myself hated tobacco or because I knew how unhealthy this was for the chimpanzee? Years later the answer is a little clearer to me: a deep empathy for these animals, forced to exist under these terrible conditions, motivated me to become a chimpanzee doctor.

SHARLOT'S STRENGTH

I had not envisioned that, thirteen years after my trip to Entebbe Zoo's chimpanzee exhibit, I would be a doctor of these very animals, fighting for their survival in the jungles of Uganda. However, many years later I was recruited by the Jane Goodall Institute in Uganda to train as a veterinary intern.

Budongo Conservation Field Station is located high atop the Albertine Rift in the moist, semideciduous, and largest tropical rainforest in Uganda. It was founded by Vernon Reynolds in 1990 following his studies of wild chimpanzees over the previous three decades. The field station had called upon veterinarians from the Jane Goodall Institute to help them rescue a chimpanzee who had been orphaned in the Budongo forest. The young chimpanzee had suffered greatly when her mother was entrapped in a horrific metal-jawed trap set by poachers. Though these traps are less commonly used in the area than smaller wire that injured my favorite chimpanzee named Zig, but the impacts of the large traps are much more pronounced. In this case, the mother died, but her daughter, Sharlot, survived.

While Sharlot had survived, she faced yet more challenges. After her mother's death, Sharlot stayed close to her mother's body, but she was soon surrounded by a troop of baboons. Usually the baboons are harassed by, and frightened of, the chimpanzees, but with only this small, vulnerable chimpanzee present, perhaps this was an opportunity for revenge! Luckily for Sharlot, the main Sonso chimpanzee community was feeding nearby and when she screamed out at the sight of the baboons, it attracted the attention of the large Sonso males. We were surprised but happy to learn that they came to Sharlot's rescue and the baboons were forced to scatter.

Shortly after, Sharlot was adopted by Wilma, a maimed female in the Sonso community who had never conceived again after the death of her one son. Wilma's adoption of Sharlot was the first I had heard of, and it gave me a new perspective on how I viewed chimpanzees. They could feel the same empathy I felt and take care of an infant in need. I fell in love with chimpanzees all over again and Sharlot will always be one of my favorites because of her resilience.

Later, I passed an interview to become the head veterinarian and conservation coordinator at Budongo. It was a dream come true to work with wild chimpanzees, and I was too excited to even ask what conservation activities I was to coordinate!

Zig

FINDING THE BALANCE

As veterinarians, we are trained how to manage the health of captive animals that live under human care, but we get minimal exposure to wild animals—even in Uganda, which has so much wildlife. As the conservation coordinator for Budongo, however, I discovered that the health and well-being of the wild chimpanzees was very much interrelated with forest-edge community health and well-being. On one hand, I was seeing chimpanzees that were injured and maimed due to human activities, while on the other hand, I was seeing important agricultural crops damaged by raiding chimpanzees. Seeing both sides of this conflict made me want to do more to reduce human-chimpanzee conflict so that both species could thrive. It became clear that only addressing one of those two viewpoints would not be successful.

I can recall a good example to illustrate the need for this balance. One day, we were alerted to an incident in which a chimpanzee was raiding the crops of a cocoa farmer near Budongo, and it resulted in a fight between them in which both the chimpanzee and the farmer were injured. Such conflicts are not uncommon in this area. In the Bunyoro area, where Budongo Forest is located, chimpanzees are known as "ebitera," meaning "the things that beat us," so, unsurprisingly, there is a lot of fear associated with the species.

The veterinary teams from the Budongo Conservation Field Station, the Jane Goodall Institute, and the Ngamba Island Chimpanzee Sanctuary all arrived quickly in three vehicles. A large group of us rushed into the village and frantically asked where the chimpanzee that was injured was. The villagers stood and stared at us in shock and then walked away.

We did not realize what we had done at the time, but the local leaders later scolded us for having no hearts. They said, "Your chimpanzee attacked one of us and you come here asking for your injured chimpanzee without any concern for the human, if he is dead or alive. Three cars have all come for one chimpanzee? The other day you people came with blankets for burying your dead chimpanzees yet here we are with no blankets to cover humans! Are you suggesting that chimpanzees are more important than us?" These questions changed my approach to chimpanzee conservation in many ways. I realized that if we did not involve local communities and demonstrate that their lives are as important as the chimpanzees', we would not achieve our conservation goals.

POVERTY PREVENTS PRESERVATION

Chimpanzees, like so many other species, face a wide range of challenges in their struggle to survive. Those challenges are complex and depend greatly on locality. Chimpanzees in Uganda face different challenges from those living in Central or West Africa. Even within Uganda, chimpanzees living in forest reserves face different challenges compared to those roaming within the boundaries of a national park. From my experience in Uganda, which is home to more than five thousand wild chimpanzees, the most apparent challenges to chimpanzee conservation are habitat loss, poaching, and disease. However, if I conduct a root-cause analysis, it seems the underlying problem can be identified as human poverty.

In Uganda, poverty is attributed to the growing human population, which exerts increasing pressure on the chimpanzee habitats. Uganda has a 3.3 percent annual population growth rate, ranking among the top ten growing populations in all of Africa. A lack of resources and income has driven girls out of school, forcing them to get married at a young age and leading to early pregnancies. The rapidly growing population means increased demand for agricultural land, which catalyzes habitat loss and degradation. With this conversion of wild lands to harvested lands comes increased contact between humans and chimpanzees. That contact begets conflict as well as an amplified risk of disease transmission between the two related cousin species.

It is clear that in order to address the apparent challenges for chimpanzee conservation, the issue of human poverty must be addressed—poverty both of resources and of knowledge. There is no single solution to the challenges facing chimpanzee conservation because the situation is so complex. But this can be overcome with a multidisciplinary approach. Academics need to work closely with the forest-edge communities in addressing the problem of poverty and the challenges of community vulnerability and resource deprivation. People need to be trained how to successfully implement intensive agriculture on their small pieces of land. Accessible education will help empower women and girls, reducing the likelihood of early pregnancies and thus reducing the human population growth rate. As effective social services become more affordable, we can educate communities on the risks of wildlife consumption. All of these changes work to address deficiencies and challenges in the human communities but also ultimately benefit chimpanzee conservation.

REFLECTING ON THE PAST TO BUILD
A NEW FUTURE

It is amazing to think that as a child growing up in the southwestern part of Uganda, I had never seen nor heard of chimpanzees in our area despite hundreds of chimpanzees living just a few miles away in Bwindi Impenetrable National Park. This may seem shocking, especially to those who would travel thousands of miles around the globe to visit chimpanzees as part of ecotourism, but even today, many Ugandans have never seen any of the wildlife that lives throughout our beautiful parks. In fact, recently my mother came to visit me and I decided to take her to the zoo where I first saw chimpanzees. When we finished our visit, she told me that the animal that amused her the most was the black one that looked and behaved like a human being. She had no idea he was the very animal that her own daughter had dedicated over ten years to conserving!

Chimpanzees have been a sort of gift to me. They have influenced my attitude toward conservation of nature. The empathy that they show is what has turned me into an empathetic conservationist. They have inspired me to work with children, teaching them mitigation measures toward environmental degradation. Although this path seems to be quite different from what veterinarians are known to do in my country, I still find myself working passionately to ensure a safe environment for all species. Seeing the instances of maimed chimpanzees injured in human-set traps, the impact of their increased respiratory infections, and the devastation of chimpanzee habitats happening in my own country, I feel like the chimpanzees need my help now more than ever.

I am Ugandan, I am a veterinarian, and I am a scientist, but most of all I am a compassionate human looking for ways to help my species best coexist with chimpanzees.

6

ANNE PUSEY

Dr. Anne Pusey was director of the Jane Goodall Institute Research Center and is a professor emerita of evolutionary anthropology at Duke University. She has been studying the chimpanzees at Gombe National Park in Tanzania since the 1970s, and her research focuses on social evolution and chimpanzee development. Although a renowned primatologist, she also studied the behavior and sociality of lions in the Serengeti for over a decade.

BEGINNINGS

My chance to study chimpanzees came unexpectedly. Like many of the contributors to this volume, I was interested in animals from a very young age. Inspired by a primary school teacher, I watched birds and kept nature diaries. My friend and I collected pond life and examined these creatures under a microscope my father brought home. Opportunities to watch wild mammals in England were limited, but I read books like *Born Free* by Joy Adamson and watched all the nature programs that I could. I had a pet cat and kept a pony for a while. Because my father, a comparative anatomist, worked in the Department of Zoology at the University of Oxford and was a colleague of the renowned ethologist Niko Tinbergen, I was aware early on that it was possible to study animal behavior as a profession. I eventually studied zoology myself at the University of Oxford. While my male undergraduate colleagues went off to Africa to study large mammals during the summers, such opportunities seemed out of reach for women. The director of the Serengeti Wildlife Research Centre at the time, when asked about allowing single women to work there, was said to have remarked, "What would they do if their Land Rover broke down? And anyway, they would cause havoc with the marriages."

I was, nevertheless, determined to do a PhD. In my final year at Oxford, I explored possibilities at different universities. In my interview with Tinbergen, he agreed to take me on to study some aspect of black-headed gull behavior as long as I did well in my exams, but he dampened my enthusiasm by warning me that, in his experience, women only succeeded in research if they were working with a male scientist. I wrote to several other professors, including Robert Hinde at the University of Cambridge, to whom I expressed my interest in applying ethological methods to studying children, something that was just starting to be done at that time. Unbeknownst to me, he had been Jane Goodall's graduate adviser, and around the same time she had let him know that she was on the lookout for assistants to help with her chimpanzee study. The next thing I knew, I had a cryptic message in my college mailbox to call a number in London. When I called it, I reached Hugo van Lawick, Jane's then husband. He said, "You may have heard of us; we study chimpanzees in Tanzania." After blanking for a moment, I remembered that I had indeed heard Jane and Hugo being interviewed on a BBC radio program a few months before. I recalled how exciting, but dangerous, their study sounded as they explained how they had built a cage with doors open at both ends into which they could retreat when the chimpanzees, in the early stages of habituation, became overexcited and aggressive! Hugo told me that Jane was looking for a research assistant and invited me to come for an interview. I swallowed my trepidation and, leaping at this extraordinary and unexpected opportunity to work in Africa, prepared by reading all that I could about Jane's work.

The interview was conducted in a small apartment in London, which I later discerned was Jane's mother's flat. Crowded in the room were Jane and Hugo, and their young son, Grub, as well as Jane's mother, Hugo's mother, and one of Hugo's brothers. Scattered in piles on the floor were Hugo's amazing photographs of African animals, from which they were making selections for their new book, *Innocent Killers*, about carnivores in the Serengeti. Grub was only about three years old at the time, and as the adults sorted through these photos, and Jane discussed the job with me, I joined Grub on the floor. Whether because of my interview, or because of my ability to bond with Grub, they offered me the position. I was ecstatic and immediately after I completed my final exams, I was straight off to Tanzania to work as Jane's assistant to work at her field site in Gombe National Park.

THE TRIALS OF ADOLESCENCE

My first job was to study mothers and their infants. Jane was interested in differences between chimpanzee mothers in their maternal behavior. Because female chimpanzees at Gombe spend a lot of time alone with just their dependent offspring, I not only got to know the mothers and their infants as I followed them for hours through the forest but also the older offspring who continued to spend time with their mothers long after they were weaned. I became fascinated by the process by which young chimpanzees transitioned from complete dependence on their mother to their very different lives as adults, and I resolved to do my PhD on this process.

My thesis was eventually titled "The Physical and Social Development of Adolescent Chimpanzees." I found that males and females followed markedly different paths but that behavioral changes in both sexes were propelled by the large hormonal changes that occurred during puberty and adolescence. Although we couldn't measure hormones directly at that time, their effects were reflected in the physical changes that we could observe and measure: rapid testes growth and a growth spurt in body size in males and the onset of large sexual swellings in females.

One of the young males that I watched navigate this adolescent period was Goblin, who was a real player. He was very political and also very social. As a juvenile, he still spent considerable time with his mother and his little sister, Gremlin. But he always wanted to play with the other youngsters and interact with the adults as well. He tried to join groups that he could hear calling while his family were feeding alone. But when his mother wouldn't follow him, he looked really torn, sometimes throwing tantrums, screaming and rolling around in the dirt. Eventually, as he reached puberty, he started spending more and more time away from his mother, joining the groups of adult males and females. He was fascinated by the alpha male, Figan, and constantly followed him around, emulating his every move. When Figan charged into a group in an aggressive display, dragging branches and kicking buttress roots, young Goblin would run right after him and kick the roots as well! As Goblin continued to grow, he started displaying at the females to exert dominance over them. At first, the results were not pretty. The older females retaliated and beat him up.

My study stopped when Goblin was a ten-year-old adolescent, still following Figan everywhere, but sometimes returning to his mother for long grooming sessions. Years later, Goblin's ambition overtook his admiration, and he overthrew his idol for the alpha position, which he maintained for many years. He was not

Goblin

particularly big, but he was incredibly determined and, perhaps because of all that practice as a youngster, he was an energetic dis-player. But when he was eventually overthrown himself, he was very brutally attacked by the group and sustained life-threatening wounds from those he had terrorized with his displays through the years.

While all the males I studied as youngsters stayed in the group as adults, the females' trajectories were much more varied. Fifi, the daughter of the old high-ranking female, Flo, traveled constantly with her mother until around the age of ten, when she matured and started to get regular sexual swellings. At this point she began spending more time in social groups and mating with adult males, while still returning to the companionship of her mother. When I arrived at Gombe, Fifi was away and everyone was wondering where she had gone. She reappeared a few weeks later, and she was pregnant. Upon her return she resumed her close relationship with her mother until Flo died.

In contrast to Fifi, Gilka's experience was very different. Her mother had died when she was nine years old; when I first met her, she was a lonely young chimpanzee, with the added disadvantages of a wrist that was partially paralyzed and a bulbous swollen nose resulting from a chronic fungal infection. She spent hours alone, or sometimes with her older brother, and when she traveled with adult females, attempting to play with their infants, they were not friendly and their juvenile offspring bullied her. Things improved in some ways when she started cycling and getting sexual swellings, at which point the males took a new interest in her. When she was swollen, she would approach males and wave her beautiful pink bottom at them. Juvenile and adolescent males, and some of the adults, mated with her and also groomed her much more than when her sexual swelling was not swollen. During this period, the habituated community at Gombe was splitting into two, and the southern subgroup of males rarely visited the north where Gilka lived. One day, on one of their last visits, I noticed that Gilka was paying more attention to these males than the more familiar males of her subgroup, and when the males left for the south, she followed them and was gone for the next five months. She later returned to her natal group but died young.

While I was not able to observe Gilka's interactions in her new group, I learned a lot from following Patti, a sturdy young female who appeared in the community I was observing for the first time as a young adolescent. She then joined the community permanently when she was about ten years old. Patti encountered a lot of aggression from the resident mothers at first. But she held her own and countered their aggression by screaming loudly, which attracted the attention of adult males who came and defended her from the female attacks. Even when she did not have a sexual swelling, she spent most of her time in groups with adult males and older females, and she formed a strong bond with young Goblin. Though he sometimes displayed at her, the two would groom frequently and mate when she was swollen. Patti remained in the group and ultimately became a successful and high-ranking female.

FLEXIBILITY AND LIONS

I was totally engrossed in the lives of the chimpanzees and was expecting to continue studying them for many years, when a shocking event upended these plans and, for a while, changed the course of my career. Though very unsettling at the time, this enforced flexibility was valuable in giving me a broader perspective on animal behavior.

Toward the end of my PhD, several students were kidnapped at Gombe and, as a result, we all had to leave. Fortunately, the students were released, but Gombe was considered unsafe for foreign students for many years. I had enough data to finish writing my thesis but rather than continue studying chimpanzees, I had to go and do something else. I had met Craig Packer at Gombe; as a Stanford undergraduate, he had started studying the baboons, and we eventually got married. It was the early 1970s and new ways of thinking about animal behavior were developing under the labels of sociobiology and behavioral ecology. There were many exciting new questions in the air and a new emphasis on the importance of kinship and the puzzle of altruistic and cooperative behavior. Both Craig and I wanted to study these questions, but without Gombe we were without animal subjects to study them. We looked around for another place to work and eventually ended up taking over the Serengeti lion study that was initially started by George Schaller in 1966, which our friends David Bygott and Jeannette Hanby had been continuing for the last four years. They had habituated many prides and had maintained genealogical information on a large number of individuals. Thus, there was a lot known about each individual lion and who was related to whom in quite a few of the prides. I worked with Craig on lions for about eleven years and we were able to tackle many of the burning questions in behavioral ecology because of the depth of the data and the large number of groups we could watch.

Watching a completely different species was incredibly instructive. It is an experience that I strongly advocate for any young biologist. Patterns that we took for granted in species we had studied previously (chimpanzees and baboons), such as clear differences in dominance relations among individuals, were not present in the lions. While I was used to the rather prickly, and sometimes hostile, relationships between female chimpanzees, I found that female lions seem to love each other unconditionally. Although they squabble at kills, they maintain no personal space. When a female returns to her pridemates, she affectionately flops on top of them, they rub heads, and then they languorously lick each other under the chin. They sleep in contact and allow others' cubs to nurse. Unlike male baboons, Craig learned that male lions form lifelong bonds with their pridemates, leaving the pride in a group and taking over other prides. Those young males unfortunate enough to lack pridemates team up with single males from other prides and form bonds just as strong.

To put it in a nutshell, going off to study a completely different species was a challenge, but it gave me a whole new and comparative perspective, which undoubtedly aided me when I later returned to study chimpanzees.

THE LONG GAME

While I was in the Serengeti, I kept in touch with Jane, who had continued the daily study of the Gombe chimpanzees with the help of her experienced Tanzanian research team. I was keenly interested in how the lives of the young chimpanzees I had studied were unfolding. I knew it would take years to get a complete picture because a long-lived species like chimpanzees can live at least into their fifties and possibly their sixties in the wild. To get the whole arc of a chimpanzee's life would take six decades!

In the mid-1980s, Jane Goodall finished her magnificent book on chimpanzee behavior, *The Chimpanzees of Gombe*, and attended the first "Understanding Chimpanzees" conference in Chicago, convened to celebrate the book's publication. As she tells it, she went in as a researcher and came out as an activist working to improve the lives of chimpanzees everywhere, compelled by what she had learned about the plight of chimpanzees in Africa, in labs, and in the entertainment industry. I had recently become an assistant professor at the University of Minnesota, and because my kids were reaching school age I no longer spent time in the Serengeti. With my new access to computers and a student workforce, I offered to help Jane by making a start on computerizing the masses of field notes and data from Gombe. And so began our decades' long effort to digitize all the information collected during this incredible study so that it could be archived and made more easily searchable.

Jane and the Jane Goodall Institute continued to fund and run the Gombe research site, and I supervised countless undergraduate assistants to computerize data. I also supervised graduate students who went to Gombe to collect data of their own on many different subjects, often incorporating new techniques such as video recording and DNA analysis from feces that we could collect noninvasively.

These students were increasingly able to mine the accumulating data from previous years to augment their own data and reveal patterns not discernible from one or two years of study.

Long-term studies, though enormously challenging to maintain, are the only way to gain insight into the social systems of long-lived animals like chimpanzees. One of the early discoveries that emerged from the Gombe studies of chimpanzees and baboons, and of the Serengeti lions, was that in all these species, the members of one sex stay in the group they were born in, while most or all members of the other sex leave their natal group as subadults and spend the remainder of their life in new groups. Craig Packer found that all female baboons at Gombe stayed home, while all males left. I showed that male chimpanzees (like Goblin) remained home while most, but not all, females left—a pattern also described at the same time by Japanese researchers studying chimpanzees at another long-term study site in Tanzania. Together, Craig and I documented the dispersal patterns of lions, where females stay home and males leave. Similar patterns started emerging from long-term studies of other animals including birds, where females are usually the dispersing sex. These patterns have profound consequences for the nature of social relationships in groups. The sex that stays home tends to be more social and cooperative, except in the case of lions, where males leave together with their kin and continue to cooperate with each other as much as females do. We argued that this sex-biased dispersal is a mechanism to avoid inbreeding.

We found that, rather than being driven out by residents as they reached adolescence, young male baboons and female chimpanzees seemed to have a spontaneous drive to leave and explore new groups and especially new mates, even in the face of aggression from members of the new group. Nevertheless, not all female chimpanzees at Gombe left their groups permanently. A major reason motivating me to return to Gombe was to determine why. Now, after more than thirty years, we have accumulated the life stories of enough females to have some answers. Females with more maternal brothers in the community, who pose a particularly severe risk of inbreeding, are very likely to leave. But females whose mothers are still alive when they reach adolescence sometimes stay home, particularly if their mothers (like Fifi's) are high ranking. These daughters continue to share their mother's feeding areas and benefit from her social support.

LIFE LESSONS

My experiences in watching chimpanzees' lives unfold have given me perspective on my own life. In the spontaneous journeys of young females to new communities, I recognize my own excitement in leaving home for new adventures, first at Gombe, and then at Stanford, on yet another continent, for my graduate work. I also resonate with the challenges that such upheavals bring, in terms of losing the close social support of relatives and friends, learning how to live safely in new places, and forging new relationships. In raising my own children, I found it comforting to know that their tantrums were not so much the consequence of my bad mothering as inevitable conflicts also experienced by chimpanzee mothers and their increasingly independent but demanding infants. Our work on the chimpanzees shows the enduring importance of mothers. Youngsters whose mothers die before they reach the age of ten suffer long-term deficits and die early. This effect continues to age fifteen for males but not females. Even so, males, and those females that remain home, maintain lifelong relationships with their mothers. Uniquely, in humans, even children who move far from their families can maintain these supportive bonds. I exchanged blue aerograms regularly with my mother while living in Africa and the United States, and texts and social media keep me in even more constant contact with my daughter and son, living at opposite ends of the country. I rejoice rather than gripe over my opportunities to provide these young adults with continuing emotional (and sometimes material) support.

7

TETSURO MATSUZAWA

Dr. Tetsuro Matsuzawa is internationally known for his research on chimpanzee intelligence, which he conducted both in laboratory and field settings, and he was formerly director of the Primate Research Institute of Kyoto University in Japan. He complemented innovative touchscreen-based investigations of chimpanzee memory and cognition there with field experiments conducted with a group of wild chimpanzees in the West African country of Guinea.

A CHERISHED CHILDHOOD

I was born in Matsuyama, Japan, in October 1950. Matsuyama is located on Shikoku Island. I was too young to remember my days living in Matsuyama, but I have in my possession a photograph album filled by my parents. One of the earliest photos shows a young boy climbing a tree. The boy actually clings to the tree trunk by his arms and legs and looks like a little chimpanzee. Underneath the photo a caption is written in my father's handwriting: "Tetsuro at two years old." The style of the script reflects his personality: meticulous and honest.

My oldest memory dates from my time in Tokyo, where I shared a small room in a downtown apartment with my parents and two older brothers. My parents had decided to move to the capital so that their three children could receive a good education with improved future prospects. After World War II, the country was in the process of recovering and had shed the status of occupied Japan to regain independence in 1952.

My parents were both primary school teachers. They worked hard and built their own house in a suburb of Tokyo. We had a small dining room where the family gathered every evening to eat and chat together. My father would say, "Today, my kids did such and such a thing," and my mother would respond with, "Today, my kids did another such thing." Their discussion of "my kids" referred not to us, their real children, but to those they taught at their respective schools. The topics of discussion at dinner were always what had happened where they worked and the process of education. This was the atmosphere of a modest family in those days in Japan.

I went to state schools in Tokyo from kindergarten through high school. Ryogoku High School was one of the best in Tokyo at the time. Most of my classmates wanted to go to the University of Tokyo, the top university for various disciplines, where many leading figures had been educated. I was naive and followed their view. However, my motivation to do so differed somewhat. I also sought to reduce my parents' burden. At that time, monthly tuition for a national university was 1000 yen, less than 3 U.S. dollars per month, which was cheap even for that era. I wanted to go to a national university, and the nearest happened to be the University of Tokyo.

THE CHALLENGE OF THE SUMMIT

In January 1969, when I was eighteen years old, the University of Tokyo announced that no entrance examinations would be held that year. This was due to the worldwide "student power" movement.

Young students protested against the establishment in major cities around the world such as Paris, London, and Tokyo. Most of my classmates then decided to go to Kyoto University, the second-best university, and so did I. Kyoto University was not strong in law and economics but had a great reputation for science, having produced the first Japanese Nobel Prize winner, Hideki Yukawa, in 1949 for theoretical physics.

Kyoto University held entrance examinations, but the situation in terms of student protests was similar to that at the University of Tokyo. Since the university staff was locked out by the protesting students, there were no classes and no teaching took place. I had no other choice but to spend my time engaged with mountaineering-club activities. Thanks to my older brother, I had begun to climb mountains during high school. Kyoto University gradually regained order and resumed teaching. However, my mountain-climbing activities continued. I spent about 120 days on climbing expeditions to various mountains across Japan spanning four seasons. A further 120 days were taken up with pre-expedition training, such as running through the old city and day-long rock-climbing trips. I spent the remaining 120 days studying at university. Whenever I was asked, "What is your major?" my answer was always "My major is mountaineering!"

Kyoto University Alpine Club, with the Academic Alpine Club of Kyoto for alumni, has an established tradition for first ascents of Himalayan mountains. The alumni count among their number several renowned scholars, including Kinji Imanishi, now considered the father of primatology in Japan. Imanishi passed away in 1992, and I belong to the last generation who met him in person. When I was a student in my twenties, he was an academic in his seventies. The founding motto of the Alpine Club was a pioneering spirit to attempt first ascents. This same pioneering spirit toward climbing also penetrated academic fields. Imanishi and his colleagues sought to establish brand-new disciplines. For example, Imanishi was the initiator of fieldwork with wild Japanese monkeys on Koshima Island, which he started in 1948. In 1958, he carried out the first survey of wild gorillas and chimpanzees with his student Junichiro Itani. Imanishi was also the founding editor of the scientific journal *Primates*, the longest-running English-language journal in the field of primatology and of which I later became editor.

During a series of mountain-climbing expeditions, I learned gradually from my fellow teammates how to become a pioneer myself—by being all around and complete in my approach. To that end, I had to work step-by-step, one by one, day by day. I came to establish my own perspective on the question "How do we see the world?"

THE WORLD AS VIEWED BY ANIMALS

My actual major at the undergraduate level was philosophy. I wanted to learn philosophy at university because this topic appeared to cover all disciplines. I enjoyed biology, chemistry, physics, and also history, geography, logic, and languages, and so I did not wish to focus on only a single subject. My dream was to have a life somewhat similar to that of the Greek philosophers. In my imagination, Aristotle and his colleagues, all robed in white, would promenade, viewing nature and discussing various topics of deep import.

When I was a young philosophy student, an old professor once told me that philosophy has two major missions. The first mission is to know the world. You must learn all about the various aspects of the world covered by the natural and social sciences. The second mission is to demonstrate the principles of how to behave in this world. Even if you understand this world completely, how to behave in it is an altogether different question. In sum, through understanding this world you have to come to understand the right way to behave in it. It seems to me that there is a third mission in philosophy: the mission to know ourselves, as humans. This world is perceived through our own perspective. It is human nature to filter the physical world. Therefore, you have to know how to actually perceive the physical world. In other words, understanding human nature must represent the third and newest mission of philosophy.

As a student of philosophy, my favorite books were about cyclopean eyes and random-dot stereograms (akin to the Magic Eye images that gained popularity in the 1990s), which discussed the idea of the split brain. This reading led me to learn more about human binocular vision at the undergraduate level and physiological analyses of the rat brain at the graduate level. I acquired techniques to observe and analyze rat behavior and also the split-brain technique to understand hemispheric asymmetry. As a graduate student, I spent almost two and a half years exploring possible left-right brain asymmetry in rats. Through the study of rat brains and behavior, I found no evidence of asymmetry and determined that knowing the rat brain did not equal knowing the human brain. Furthermore, I became convinced that the brain is a part of the body while the body is a part of the environment.

I shifted from an analytical approach to a more holistic perspective. This change was supported by my favorite books at that time, all on ethology, including Nobel Prize winner Konrad Lorenz's *King Solomon's Ring* . However, the idea that had the most influence on me was that of *Umwelt*, which was presented in the 1926 book by Jakob Johann von Uexküll. The book was translated into Japanese by Toshitaka Hidaka, a contemporary Japanese ethologist, and the title was *The World as Viewed by Animals*. This concept is woven throughout my life's work.

PARALLEL PATHS

In 1976, a vacant post at the Primate Research Institute at Kyoto University was announced. It was an assistant professorship within the Department of Psychology. At the time, I was a PhD student in the middle of my third year. My studies stemmed from this philosophical question: "What is human nature?" While my undergraduate work was on human vision, especially binocular depth perception, my graduate research involved controlling the behavior of rats to understand their brain function. Uniting all these topics of investigation, I wrote a proposal for the advertised job. My aim was to understand the perceptual world of nonhuman primates through behavioral observation and cognitive experiments. I believed that studying the nonhuman primate mind represented a unique way to understand human nature. At the age of twenty-six, I was very lucky to get a tenured position.

One year later, in November 1977, a one-year-old female chimpanzee named Ai arrived at the Primate Research Institute. This marked the start of the Ai project. The Ai project was unique because it was a fully automated study focusing on psychophysical measurement of chimpanzees' perceptual and cognitive functions. In the laboratory at the institute, after the cognitive experiments ended for the day, I would get into the same booth as the chimpanzees to interact with them directly. I would emit grunting noises to greet them, and I would perform the "hug" and "open-mouth kiss" gestures that chimpanzees did with each other. I prefer to interact with chimpanzees using their own natural vocal and gestural repertoire. My pant-hoot calls work: the chimpanzees always respond to my pant-hoots, both in the laboratory and in the wild!

In a parallel effort to my work at the institute, I started my fieldwork on wild chimpanzees in February 1986, focusing on stone tool use by the chimpanzees of Bossou, Guinea, in West Africa. There

Ai

is a sort of cultural heritage shared by the chimpanzees unique to Bossou. There, chimpanzees use a pair of stones as a hammer and anvil to crack open oil-palm nuts. There are other kinds of tool use and manufacture: folding leaves for drinking water, pestle-pounding of the top of oil-palms, and stick use for scooping out algae floating in ponds. I would follow the chimpanzees all day, every day. I would get up at five thirty in the morning, entering the forest by six thirty. I would find the chimpanzees still asleep in their nests and follow them throughout the day until six thirty in the evening, when they made their night-nests and emitted soft grunts before sleeping. I would sample what they ate: fruits, young leaves, stems, bark, gum, underground roots and tubers, nut kernels, floating algae, ants, termites, etc. The only thing they ate that I never did was the flesh of pangolins.

Putting together observations from the field and the laboratory, I could go back and forth to understand the chimpanzee as a whole. Sometimes I almost wish I could turn into a chimpanzee to see directly their experiences. Through the eyes of a chimpanzee, I would seek to understand the perceptual world of chimpanzees and thus to know what is uniquely human.

JOKRO

The greatest challenges facing wild chimpanzees today are habitat loss, illegal hunting, and contagious disease. All of them come from human activity. Although there is no hunting at Bossou, we witnessed young chimpanzees trapped by poachers' illegal wire snares, which led to them losing fingers or a hand. The number of chimpanzees there continues to decrease.

I spent three months alone at Bossou in 1992, and I saw an infant chimpanzee die of a flu-like disease. Her name was Jokro, and she was two and a half years old when her life ended. Her mother, Jire, was approximately thirty-five. I videotaped the final sixteen days of Jokro's life and the twenty-seven days thereafter. And so an infant chimpanzee lived and died in the forest. Her life was short, but I observed how she remained with her community for a month beyond her death as her mother continued to carry her body. The episode remained deep in my memory. The word *Jokro* comes from the language of the area's native people, the Manon. It is the word for the triplochiton, a huge tree with leaves in the shape of a hand. I hope Jokro will be reborn in her next life as her namesake and live for a hundred years.

Chimpanzees need those trees like the Jokro. Many years ago, I started a conservation program to plant trees in the savanna, which we called the Green Corridor project. It aims to make a greenbelt of trees connecting the isolated forest of Bossou with the neighboring Nimba Mountains, a UNESCO World Heritage site where a thriving community of hundreds of chimpanzees live. Learning from the wisdom of the local people, we started utilizing the feces of chimpanzees in the planting because the grains inside the feces are good for germination. The young trees in the tree nursery will provide the fruits consumed by the chimpanzees and in turn, the chimpanzees are in charge of the seed dispersal, which supports the biodiversity of the tropical forest. The trees need the chimpanzees, just like the chimpanzees need the trees.

REFLECTING BACK

There have been three key events in my life as a chimpanzee researcher. First, I happened to go to Kyoto University, where I became involved in mountain climbing and pioneering work. Second, I gained a position at the Primate Research Institute, which provided me with a secure base to support my research for more than four decades. Third, I met Ai the chimpanzee. After studying her intellectual capability, I went to Africa. I wanted to visit where she was born—in other words, where the chimpanzee mind was shaped through evolution. She became both my research partner and my navigator and helped me reach into the world of chimpanzees.

However, the photo of me climbing a tree at two years old seems to convey that things take place as if predetermined. The Manon people of Bossou respect chimpanzees as their ancestors. Neighboring tribes believe that the Manon turn into chimpanzees in the forest: when the chimpanzees come out of the forest into the village, they transform back into Manon people. I write this essay at the end of my twenty-eighth annual survey of the Bossou chimpanzees. Just like the Manon people, at the end of my career, my dream is to turn into a chimpanzee living in the forest.

8

CHRISTOPHE BOESCH

Dr. Christophe Boesch was director of the Department of Primatology of the Max Planck Institute for Evolutionary Anthropology in Leipzig, Germany, and is active as the president of the Wild Chimpanzee Foundation, which runs conservation programs in Côte d'Ivoire, Liberia, and Guinea. In 1976, after conducting census work on the mountain gorillas of Virunga National Park in Rwanda, he began studying the chimpanzees of Taï National Park in Côte d'Ivoire. The park is home to eleven primate species. The chimpanzees there are well known for their expert nut-cracking culture.

A CHANCE ENCOUNTER IN A BOOKSHOP

Chance is an important component of what determines one's life course. When I was twelve, I showed relatively little interest in animals and yet my father offered me a book by Konrad Lorenz entitled *King Solomon's Ring*. The book, which was first published in 1949, is a classic among animal behaviorists, and its title refers to a legendary ring donned by King Solomon of Israel that supposedly gave him the ability to communicate with nonhuman animals. Lorenz claims he achieved similar powers of interspecies communication through his careful exploratory studies of animals, and the book describes these studies in compelling detail. Reading this book put me on the path to study animal behavior for virtually the rest of my life.

Later, I enlisted in the scientific section of my school in Paris to study biology. A second stroke of luck occurred during this period of my life when, by chance, I was visiting a bookstore and came across a copy of *The Year of the Gorilla*. George Schaller's vivid descriptions of his time among the gorillas of the Virunga volcanoes was impactful reading for me, and I was drawn into his accounts of these powerful and majestic creatures. The random discovery of this book on the shelves of that bookstore in Geneva helped to further guide my life's work. After reading it, I said to myself, "*That* is what I want to do!" And so I did.

I enlisted in the biology program at the University of Geneva in Switzerland to continue my training, but it was not until I arrived there that I realized that my ethology professor knew Dian Fossey personally from her time in Cambridge. It was yet another stroke of good fortune that would shape my career path. Through her, I worked to organize my master's thesis with Dian Fossey, who was studying the mountain gorillas in what was then called Zaïre (now the Democratic Republic of Congo, DRC). She was organizing a complete survey of that population and needed a French-speaking person to communicate with the authorities on the Congo side.

Before I knew it, I found myself on the steep slopes of the volcanoes of the Virunga Mountains tracking the majestic and impressive gorillas in one of the most magnificent, luscious forests of Rwanda. This key experience convinced me to study wild animals in Africa, but even then, I knew that I needed to have my own field site to conduct my studies.

FORGING MY OWN PATH

Starting a brand-new field site to study wild animals is no small task. But again, through a piece of luck, I was presented with a fantastic opportunity. While meeting with a colleague at a café in Paris I learned of a population of chimpanzees in Western Africa who supposedly had been seen using stone tools to crack open nuts to consume them. There were a few published reports that examined this possibility but no direct observations to confirm the behavior. I could not pass up this exciting prospect! Thanks to a little financial support, in the mid-1970s I found myself traveling to the Taï Forest National Park in Côte d'Ivoire. I arrived with virtually no knowledge of the chimpanzees there. With only an old, dying Land Cruiser and a shoddy canvas tent, I was determined to find proof that the chimpanzees were the legendary nut-crackers, carefully toiling in these dense forests.

Observing any animals that consider humans a threat is a challenge, and I quickly came to realize that these chimpanzees would invariably scurry away from me upon my arrival in the forest. Such challenges did not dampen my enthusiasm; I was intent on discovering a potential unknown tool use in West African chimpanzees. I spent month after month in the forest, but most of that time was outside the season in which the nuts were edible and attractive to the chimpanzees. But when the nuts started to ripen, I grew anxious and spent hours waiting to see what the chimpanzees would do.

Finally, one day I was drawn to the staccato sounds of hammering deep in the forest. I spotted a female chimpanzee acting a little differently from what I was accustomed to seeing. I carefully and quietly peered through the foliage and spied her selecting a stone from the ground, preparing to crack open a Panda nut. This was it! However, she quickly realized I was watching her and vanished into the forest before striking the nut, but that brief sighting was the moment I had waited for since hearing about this behavior back in Paris.

This single moment was more than enough to motivate me, and I reported it back to my then adviser, Hans Kummer. We decided to begin a long-term study of the Taï chimpanzees. As I said at the outset, chance can play a large role in determining our life path. In my case, two books and that one fleeting moment in the African forest have led me to what has become a forty-year study of this fascinating population of chimpanzees.

THE TWO SIDES OF BRUTUS

Brutus was the dominant male of the community when we—my wife, Hedwige, and I—tried to habituate the chimpanzees to observe their behavior. Despite being the alpha male, he was at the same time quite careful to avoid our presence. With cuts in both ears and a distinct white beard, he looked impressive. After five years of following his group in the forest, some individuals started to tolerate us and we could then make our first direct observations. Brutus was still wary of our presence and clearly wanted to protect his group members in situations of danger.

Such danger existed in areas where chimpanzees could come into contact with humans, such as villages, crop fields, and roadways. I remember vividly watching the group cross the one dirt road leading to our camp. Brutus would walk confidently out of the forest and into the middle of the road. He would stop and carefully turn his head to check in both directions for cars and other danger. Only after he had monitored for threats would the group members scurry out of the bush and across the dusty road. Brutus would follow last—their leader and protector.

The Taï chimpanzees are expert hunters of the arboreal monkeys that reside in the forest, and as the habituation of the chimpanzees progressed, we had increasing opportunities to watch them in these activities. The hunt was hectic at times, but there was a sense of organization as well. Brutus was a great hunter and had the ability to not only anticipate the way his fellow hunters would react to the monkey prey movements, but he would also anticipate how the prey would be fleeing when trying to escape. In this way, he could work with others and form a perfect team to close the trap around the prey and increase the likelihood of a capture.

In a sense, Brutus understood that to capture a small monkey high up in a tree, it was important for the larger chimpanzees to hunt as a team; alone, their chance of success was very low. To do so, Brutus considered the position of both the prey and his fellow hunters and adapted his position to increase the hunt's chance of success. This did not mean that it increased his personal success. When Brutus appeared suddenly in the tree with the monkeys he was chasing, and they turned to run away, he would be left empty handed while others made the capture. However, thanks to the meat-sharing rules prevailing in Taï, he was still rewarded with meat after the capture was done.

Brutus was an expert hunter but was also the clear leader of his group and a chimpanzee with superior social savvy and surprising tenderness for such a large, intimidating individual. One day we saw him spend a lot of time with a young juvenile male named Ali. Ali's mother had recently, and quite suddenly, died, leaving him alone and without support in the group. Ali approached Brutus carefully, daring to invade his personal space. Brutus not only tolerated Ali's proximity but adopted him and protected him from other group members. Ali benefited greatly from this unique friendship as Brutus generously shared with him meat and nuts— much to the dismay of the nearby adult females who were not so happy to see this little chimpanzee receive so much of what they coveted. The group of begging females often started conflicts as a result, but Brutus, being the confident leader that he was, was always able to resolve them without compromising this generosity to little Ali.

UNEARTHING THE TRUE AND VARIED NATURE OF CHIMPANZEES

Over the course of four decades of studying the Taï chimpanzees, so much has been learned. But I am constantly struck by the fact that, although scientists have always had a fascination for our closest living relative, they also seem to be trying hard to ignore as much as possible about what they really are.

After spending five years habituating the chimpanzees to behave naturally in their social and ecological environment without any undue influence from my presence, I realized that most of what had been written about chimpanzees was inaccurate.

There was a perception that the tool-use behaviors that were being documented at different field sites were only brief and periodic in nature. At Taï, however, we found that chimpanzees use many tools on a regular basis for many tasks and for many months of the year. Tools are a part of daily chimpanzee life there.

There was a perception that chimpanzees were broadly vegetarian or, at best, that any meat consumption was extremely rare. But we observed that forest chimpanzees are keen hunters during four months of the year and eat insects on a daily basis.

There was a perception that chimpanzees existed without fear of predation. But we found that leopards are a very serious threat, and chimpanzees have developed some important within-group solidarity responses to fight back against them.

There was a perception that chimpanzees were almost completely engaged with competing among each other. We now know that chimpanzees actually possess a very developed sense of group cohesion and support one another in the face of danger, cooperating to hunt for meat, helping injured individuals, and even sharing food generously with members of their group.

It is important to remember that when scientists started to study chimpanzees in their natural environment in the early 1960s, it was only with two small groups in one country, Tanzania. We know now that chimpanzees exist throughout West and Central Africa in dozens of communities, each with unique characteristics and customs. So the initial set of studies was only a tiny sample size compared to the many hundreds of human societies we know and have studied for decades or more. In the last forty years, we have been able to add many more new chimpanzee groups to this sample, but we are still far removed in our chimpanzee studies compared to what we know about humans. In addition, we know that with each new chimpanzee population, we discover new behavior patterns and capacities in this species; therefore, there is a real need to study them before it is too late.

That was the main reason we started a new study in 2006 on the little-studied chimpanzees in Gabon, in Loango National Park. The Central African chimpanzee population is by far the largest in Africa, but only a couple of populations are presently under study. We used remote camera trapping to film the chimpanzees before they were habituated to human observers. In this way, we could document their fascinating behavior, such as the use of different sticks to access the honey of bees' nests that were underground. The habituation took even longer than the five years I had invested in familiarizing the Taï chimpanzees with my presence, but my patience proved to be worthwhile. Today, the Loango chimpanzees are showing us behaviors we never knew to be in the repertoire of the species. They have unique interactions with gorillas living in the same forest and, amazingly, we have observed a completely novel way of supplementing the need for protein in their diets, as they have learned to smash tortoises on rocks to exploit their meat. These and other revelations have come from the long hours of observing chimpanzees, and I am grateful to have been witness to their true nature.

The process of revealing these inherent truths about the chimpanzees has reinforced my belief that you should believe in what you see rather than sticking just to what you have read. That lesson—what the Taï chimpanzees have taught me over the years—has accompanied me throughout my life, in my research as well as in my daily life. We are often pressed to believe mainstream opinions, but it is important to remember that we are thinking animals, and we should use that ability to remain truthful to our own humanity—to not hesitate to question what we are told. The discoveries we made at Taï have threatened some of the established convictions, but as a scientist and as an animal lover, I have always felt it is absolutely the way we need to proceed to understand chimpanzees.

CURIOSITY, SKEPTICISM, AND THE TRUTH

Science should be about curiosity. Curiosity about novelty, about the unknown, about the unexpected. So I thought our discoveries of Taï chimpanzees hunting cooperatively in groups, using more tools in the forest than was known, modifying them in new ways, and of mothers teaching their young how to use tools would be welcomed by the scientific community. In doing so I underestimated the inertia of established scientists who are not so open to see their own work being challenged. Sadly, many in the fields of psychology and anthropology reacted with strong skepticism to my observations of the Taï chimpanzees; some went as far as claiming I was wrong, overinterpreting my data, or just reporting lucky anecdotes.

When I reported our first observations of chimpanzee cooperation, some saw this as a threat to one of the proposed important tenets of humanity, our unique ability to cooperate. Michael Tomasello, a colleague of mine working in the same institute, attacked my observations by regularly proposing new reinterpretations of the Taï data that would be compatible with his idea of human superiority. Amazingly, these attacks continued even after new observations in Taï and other chimpanzee populations confirmed their ability to cooperate with other group members. But because this contradicted what Tomasello and his team had observed in numerous experiments run with captive chimpanzees, and so some attempted to dismiss my findings. To them, observations on captive animals should be valued more seriously than those on wild animals because of the degree of control that scientists can have on particular variables of interest, while discrepancies with wild observations are rarely discussed.

Similarly, when I gave reports about teaching interactions between mothers and their offspring when learning to use tools to crack nuts, it was met with great skepticism. Many psychologists rejected these reports as unsupported claims because they had not been able to show this same behavior in captive chimpanzees or simply because it did not fit with their view of the uniqueness of humans, in which teaching was a key element. None of these skeptics seemed to be able to consider that wild chimpanzee mothers might have more and additional incentives to teach their offspring than a chimpanzee mother in captivity where their basic needs are met. In the years following those initial observations, we continue to document new observations at Taï and in other chimpanzee populations that support the notion of teaching by chimpanzees, but I suspect the resistance to these ideas will continue.

The last example of resistance to new ideas about chimpanzee behavior was when we first published my observations of female chimpanzees being better nut-crackers than males. At that time, a friend told us teasingly that we should not publish such things since male scientists would not like it if we contested male dominance in science. It was a funny comment then, but I fear now, after forty years of research, that he was right to some degree. Some scientists still prefer to believe results supporting superiority in human intelligence rather than wanting to learn about the fascinating behaviors of chimpanzees. Desmond Morris was already aware of this when he wrote, back in 1967, that humans are like the "new rich that do not want to be remembered about their origins." His wisdom remains very accurate.

OVERCOMING BARRIERS TO SURVIVAL

When I first traveled to the Taï Forest with my wife, Hedwige, we drove for 150 km through a seemingly endless lush green tunnel of trees. Along the way, we encountered all sorts of wildlife, including elephants and chimpanzees, crossing the narrow dirt road leading into the park. Today, things look very different. You need to drive all the way to the actual entrance to the park before you see any intact forest or any variety of wildlife. All the vast forests outside of the park are gone and have been converted into sprawling coffee, cocoa, hevea, and oil-palm plantations. Taï National Park has become an island in the middle of an ocean of bustling plantations. Over the last few decades, Côte d'Ivoire has experienced very successful economic development based on cash-crop production, but this came with extraordinary costs for the country's forests and wildlife.

In the early days of our work at the research camp, we regularly heard the noise of distant chainsaws and bulldozers from inside the forest. Even within these protected areas, the chimpanzees and their forest habitats were subject to the attacks of money-hungry corporations. Eventually, I needed to meet with the minister of the environment, waters, and forests for the park limits to be respected. If we, the researchers, had not been present, the loggers could have easily cut through the forest with impunity. Indeed, that is happening in many places of this park and other parks throughout Africa.

My experience in Côte d'Ivoire convinced me that without the legal protection of a national park, the fate of chimpanzees will be very hard to secure. Even within the boundary of a national park, effective management was necessary. So, despite my obligations as a researcher, in 2000 I decided to create a nongovernmental organization devoted to the protection of the wild chimpanzee population in Africa: the Wild Chimpanzee Foundation. In the last twenty years, and thanks to the trust of many supporters of our work, we are now well established in Côte d'Ivoire, Liberia, and Guinea. We are in the process of creating three new national parks with healthy chimpanzee populations and helping to manage two others. Most importantly, we are working extensively with local communities to sustainably increase their acceptance of the parks and contribute to the long-term survival of the critically endangered West African chimpanzees.

9

ANDREW WHITEN

Dr. Andrew Whiten is an emeritus professor in the School of Psychology and Neuroscience at the University of St Andrews in Scotland, and he is best known for his research on primate social cognition and culture. By studying both wild and captive primates, his work has sought to elucidate how monkeys and apes learn from one another and maintain traditions. He was the founding director of the university's Living Links Research Centre in Edinburgh Zoo, where visitors can observe scientists running cognitive research with capuchin monkeys and squirrel monkeys.

BEGINNING WITH BABOONS

I came to chimpanzees relatively late in my career—at least, compared to the majority of contributors to this volume. Instead, I began my studies of African primates through a focus on baboons, first in Senegal, then in South Africa, and later in Kenya. Baboons, of course, are not as closely related to us as are chimpanzees, but I chose to study them for many of the same reasons as those who study chimpanzees, which is namely the drive to understand the evolutionary foundations of our own human nature. Chimpanzees inform on this because of our shared ancestry of just a few million years ago; baboons instead illuminate what life is like for a large primate braving the African savannas, as our hominin ancestors did, dealing with more meager food distributions and enhanced predatory dangers compared to the more forested habitats favored by chimpanzees.

It was baboon life that kindled my interest in the cultural minds of primates, the core topic of the work I later began with chimpanzees. Along with my colleague Richard Byrne, studying how baboons made a living in the challenging habitat of the Drakensberg Mountains of South Africa, I became fascinated by how they came to acquire the sophisticated knowledge they revealed. To take just one example, we were initially at a loss to understand how they could walk several paces through the bush and then suddenly stop and manage to dig up a tiny but clearly valued corm, no bigger than a pea. As we examined their excavations, we realized that the critical cue was the tiny green stalk of the plant, barely discernable by its more vivid green color than the thousands of stems of grasses, sedges, and other plants in the square meter in which this corm was the single nutritional treasure.

How did the baboons know the secret of the tiny green stem? That each individual baboon might gain such knowledge through trial-and-error exploration seemed rather implausible. It was the behavior of developing infants that seemed to suggest the answer. I observed them intently watching their mothers digging up the corms—and indeed every aspect of her foraging, especially every time she focused on a source new to the infant. The infant would then frantically pile in to explore whatever remains their mother left, which was perhaps just the husk of a corm or the shelled pods of acacia seeds.

The hypothesis, then, was that these primates acquired their sophisticated understanding of their environment to an important extent by learning from existing experts—first their mothers and then others as their social world gradually expanded. This would be their version of culture, paralleling all we humans acquire from the cultures we are born into. However, this was only one hypothesis among others. For example, perhaps the focused interest of the infants was instead all about scrounging whatever they could from whomever they happened to closely watch.

CHASING CULTURE

When faced with such dilemmas, scientists' most powerful scientific tool is the controlled experiment, and the hypothesis of "social learning" or "cultural transmission" lends itself well to such an approach. We can present some of our subjects with an opportunity to learn a secret from someone who knows it and then later test their knowledge compared to other subjects who have no benefit of an expert model to learn from. That, however, is far from straightforward to arrange in the wild. Accordingly, my next research step was to engineer such experiments with captive animals.

It was then that serendipity intruded, as the first suitable research participants I happened to locate were not baboons but chimpanzees. Contacts in Spain, Juan Carlos Gómez and Patti Teixidor, alerted me to the existence of eight juvenile chimpanzees, aged four to eight, who had been confiscated during an illegal shipment and afforded refuge in an off-exhibit enclosure at the Madrid Zoo. I well recall my introduction to these youngsters, which, as my first close encounter with our sister species, was a kind of hand-to-hand combat. "Let's go in and introduce you," Juan and Patti said, and blithely in we went. But, of course, I was a novel object to these were eight desperately curious young chimpanzees. I was immediately mobbed by all of them at once, and I quickly discovered that the only way to avoid being gnawed on all sides was to crazily tickle each hairy bundle in vigorous succession! Not exactly recommended behavior, but we knew no better then.

Later, my research student Debbie Custance, having braved a similarly memorable induction, completed our intended experiment. We had created what we called an "artificial fruit," which was designed to echo the properties of complex foods tackled by chimpanzees and other primates in the wild. It required a series of twisting, pulling, and poking actions on various bits of the "shell" to get the tasty food reward inside. Since we had so few test subjects at first (later we also worked with some chimpanzees living at the Yerkes National Primate Research Center in Atlanta), we had designed each artificial fruit so there were always two different ways to remove each defense on the shell of it. For example, one knob could be pulled and then twisted or poked and then rotated out in the opposite direction. For our study, some of the youngsters saw Debbie do it one way, and others saw her do it via the alternative method. Sure enough, the experiment confirmed acquisition of this little microcosm of culture, with youngsters tending to acquire whichever of the two methods they had witnessed Debbie show them. She also tested young children in the same way. The results, echoed by many further such experiments completed in my research group, might be glossed over as "children tend to copy all they see, but young chimpanzees copy, too, even if not so faithfully as do our own species."

Following these and other experiments, serendipity showed its face again. I was invited by Mike Tomasello to review all we had learned about primate culture for a special issue of a scientific journal he was editing. In it, I summarized the results of our experiments, like that described above, but I also noted the accumulating evidence suggesting variation in chimpanzees' cultural practices across Africa. Primatologists, including Jane Goodall, Bill McGrew, and Christophe Boesch, had even begun to draw up tables indicating that behavior patterns common at some locations appeared absent at others. Of course, we find the same in human communities across Africa, which is attributed to our cultural natures. Was this variation also the case for chimpanzees?

I judged the primatologists' collective conclusions as suggestive yet scientifically wanting. They were based on what had been published from different research sites, but of course not all locally common behaviors had necessarily been reported, and local behavioral *absences* are even less likely to be reported and published. To make matters worse, authors tacked different personal terminologies onto the behavior patterns they addressed, so it was difficult to discern if behaviors differed or just the researchers' names for them did.

To my surprise, when I suggested to all the leaders of the long-term chimpanzee study sites across Africa that they should instead pool all their records and collaborate on a systematic analysis of them, they all leaped at the idea. In the first phase of this study, the nine research groups listed any behaviors they suspected to be culturally variable, which gave us a list of sixty-five candidate behaviors. In the second phase, these behaviors were pruned down to a list of thirty-nine, which we classed as putative cultural variants because they met the criterion of being common in at least one community yet absent at another without any obvious ecological explanation.

The results of this effort, published as "Cultures in Chimpanzees" in the prestigious scientific journal *Nature* in 1999, elicited a landslide of coverage in the world's media. We had claimed not merely that chimpanzees exhibited traditions but that they had quite sophisticated local cultures, each defined by an array of variations spanning foraging techniques, tool use, social behavior, and even courtship gambits. I was surprised to find that these discoveries even stimulated editorials in the *New York Times* and *The Times* of

London! The first of these began by noting that "certain behavioral variations among chimpanzees could only be called cultural . . . something that was believed to be a wholly human attribute" and then went on to declare, "The case of the cultural chimpanzees is, for some people, particularly troubling because it blurs a boundary that seemed especially clear-cut, almost sacred." I was quite shocked by this response. After all, we were not claiming that chimpanzees had *human* culture—and surely no one would doubt that there is a huge gulf between the complexities of human culture and what we were describing in chimpanzees.

What is exciting to me is that we had discovered something that a century ago was undreamed of—that the transmission of numerous and varied cultural traditions shapes the lives of our closest relatives in the forests of Africa, just as it does their human neighbors.

TOOLS, TESTS, AND TRADITIONS

When we published the chimpanzee culture paper, some of our fellow scientists remained skeptical—had we really shown that behavioral variations are socially transmitted? Might they instead reflect genetic differences between chimpanzee populations or undetected environmental determinants? Clearly, further rigorous experiments were called for to answer such concerns. However, such experiments now needed to go beyond showing that one chimpanzee may learn from observing the skills of another. Culture is a population-level phenomenon, so now we needed to look at social transmission across whole groups.

We first engineered appropriate tests in the three groups studied by Frans de Waal at Yerkes. Instead of artificial fruits, we now progressed to designing novel food-getting tasks that required the use of tools, again ensuring that such tools could be used in two quite different ways to access the food. For example, to release a grape from behind a blockage, a stick tool could be used either to lift the blockage out of the way or to push the blockage and grape backward and off a ledge so the grape could be retrieved. I called this task the "panpipes" (a combination of the Latin name for chimpanzees, *Pan troglodytes*, and the musical instrument constructed of adjoining pipes, which the shape of the task reflected). The idea was to train one chimpanzee in one social group to use one tool-use technique (lift) and one individual in a second group to use the alternative technique (push). Then we would reunite each expert with her group to see if their groupmates would watch and copy their newly found skill and to see if the techniques began to form different incipient traditions in each group.

In these and later studies, we strived to make the activities involved correspond to the kinds of tool use observed in the wild so that our chimpanzees' minds would engage with the tasks in as natural a way as possible. As the scientist who would run our first such experiment at Yerkes, Vicky Horner, put it, the set-up needed to be appropriately "chimpy."

Vicky and I had learned this lesson well when, as my PhD student, she conducted other experiments on social learning in chimpanzees rescued from the bushmeat trade and living in a wonderful sanctuary on Ngamba Island in Lake Victoria, Uganda. To probe just what differences may exist between human and chimpanzee social learning, we arranged that some chimpanzees would see Vicky first open a hole in the top of an opaque box and stab a stick into it several times and then she would open a second hatch on the side, stab the stick in there, and extract a piece of food to be shared with the observer chimpanzee. Other chimpanzees, however, saw Vicky perform the same sequence of actions, except the box was transparent, so when Vicky rammed the stick in the top, observers could see that it uselessly banged on a false ceiling inside the box—so this first action was irrelevant in obtaining the food. Later, we also presented both these boxes to children, and Vicky gave the same demonstrations as the chimpanzees had been shown.

We anticipated that both species, when allowed their turn with the opaque box, might copy all the actions they had seen, as they would be unable to differentiate relevant from irrelevant actions. With the transparent box, however, we speculated that children, as representatives of the more intelligent species, would display mental flexibility and omit the first irrelevant action of poking the tool in the top hole, whereas chimpanzees would perhaps more mindlessly continue to "ape" all the actions. What we found surprised us. It was the reverse of what was predicted. With the transparent box, the young chimpanzees were the ones who sensibly tended to omit the first action (poking in the top hole onto the false ceiling), and it was the children who continued to replicate it! It is not uncommon for these results to elicit reactions like "Wait . . . which is meant to be the smart species?" But perhaps children's blanket copying somehow reflects the all-pervading importance of cultural learning in the human species. Research to make sense of such intriguing findings continues.

Working later with the panpipes task and the Yerkes chimpanzees, Vicky, Frans, and I discovered that two different incipient traditions did indeed spread in the groups in which we had "seeded" alternative tool-use techniques through a single trained chimpanzee. A third group, with no trained model, tried but never managed to apply a stick tool in a way smart enough to extract a grape from the panpipes. Later experiments of this kind in a large research center in Bastrop, Texas, showed such traditions could even spread from group to group where the observation conditions allowed it, which corresponded to what we believe happens in chimpanzee traditions in the African wild.

MORE SERENDIPITY

It has been a privilege to collaborate with so many gifted primatologists, both in collating the observational records from the wild
and engineering these experiments with captive chimpanzees, the
results of which converge to teach us of the surprising cultural
nature of our sister species, something we could not have anticipated even a half century ago.

There was further serendipity in 2018, when I was delighted to
be invited by the United Nations Environment Programme to participate in a workshop to explore what had begun to be suspected
are important implications of the cultural lives of animals for conservation policies and practices. A thirty-page report ensued, its
recommendations summarized in an article in *Science* the following year. Its title carries its core message: "Animal Cultures Matter
for Conservation." If I have learned any lesson for how to pursue a
rewarding scientific career, whether in primatology or any another
subject, it reflects the fact that in my writing here I have used the
word *serendipity* more than once. One may set out to pursue one
question only to experience chance discoveries that are yet more
exciting and interesting if one can only recognize them as such.
Alternatively, one may find results that overturn one's original
hypotheses yet truly have the potential to take your science to a
higher level.

The eminent chimpanzee researcher who began the next major
field study after Jane Goodall's, the late Toshisada Nishida, was
famous for saying "Every chimpanzee is a new chimpanzee for me."
There are surely more surprises in store as a new generation of
researchers continues to explore the nature of our endlessly fascinating sister species.

10

MELISSA EMERY THOMPSON

Dr. Melissa Emery Thompson is an evolutionary anthropologist and professor at the University of New Mexico, where she co-directs the comparative human and primate physiology center. Since 2000, she has worked with the Kibale Chimpanzee Project in Kibale National Park in southwestern Uganda, where scientists have maintained daily records of approximately fifty-five chimpanzees ranging across the moist ever-green forests for over thirty years. Her work uses a variety of innovative and noninvasive methods to study how the physical and social environments of wild chimpanzees influence their physiology and behavior.

MY PATH TO CHIMPANZEES

I was an only child in a military family. We moved to a new place every couple of years, but my parents always knew they could keep me happy by taking me to the zoo, the aquarium, or a farm to watch the animals there, or simply to the pond to feed the geese. Other-wise, I could be occupied with an episode of *Mutual of Omaha's Wild Kingdom*. If it had fur or feathers (and occasionally, scales), I was interested.

While I did not have a specific fascination with primates early on, that changed when I was about twelve years old. My mother was a nurse on the night shift at a local hospital and one evening, she brought home a tall stack of old *National Geographic* magazines that someone had left behind. One of them was a 1979 issue that included a feature article by Jane Goodall entitled "Life and Death at Gombe." This was neither Goodall's first, nor most prominent, article, but it captured my attention.

The photo on the cover page was ominously captioned "Shadow of doom hangs over a mother chimpanzee as she cradles her infant, later killed and eaten by group members here eyeing the two." The piece delved into the dark side of chimpanzee lives, including murder and cannibalism, complete with intricate and horrific black-and-white drawings of chimpanzee violence by David Bygott. This was juicy stuff with obvious morbid appeal for a preteen, but it was the "centerfold" that made the biggest impression on me. It showed a family tree, with portraits of the chimpanzees along with individual histories and motivations. What played out on those pages was a multigenerational struggle between families, each led by a matriarch, not unlike a classic literary saga. Reading the article was not only the earliest spark for a career in primatology but it also likely shaped my desire to study the complex and subtle lives of female chimpanzees, who received relatively little attention in the intervening years.

It is not particularly surprising that a primatologist of my generation was first influenced by Jane Goodall. One can easily find doctors, lawyers, schoolteachers, and stay-at-home parents who have been and continue to be tremendously inspired by her work. Deciding to make a career of it was an entirely different thing, and an early impediment for me was simply knowing that it was even possible. Despite my immersion in the subject matter, it had never occurred to me that there was training and a career path for research on wild primates. I grew up in communities that were not near universities and, as such, I met very few people with doctoral degrees. Careers in basic research, especially in the behavioral sciences, simply never came up in the schools I went to.

In the end, I stumbled upon this career path almost by accident. I was fortunate enough to attend Emory University in Atlanta in the mid-1990s. Determined to pursue a preveterinary program, I quickly discovered a thriving primatology community. There was an extraordinary critical mass of primatologists working in and around Emory at the time, which offered plenty of opportunities to be involved in research with captive primates at the Yerkes National Primate Research Center. Ironically, having this huge presence of primatology around me convinced me that this would be an easier and more straightforward career path than veterinary science. This is laughably untrue, but I fortunately never had to learn this the hard way. I owe this to my undergraduate research mentors, Harold Gouzoules and the late Pat Whitten. They never planted any seeds of doubt as to whether I was good enough, and they instilled confidence in me by treating me like a professional colleague, criticisms and all. They encouraged me to apply to the best graduate programs, and thus, I found myself at Harvard University studying with renowned chimpanzee expert Richard Wrangham.

Still, I went to graduate school with the intention of studying anything but chimpanzees. "Chimpanzee people," I thought, were too hasty in making comparisons between chimpanzees and humans. Chimpanzee research was oversensationalized. And chimpanzees were way too complicated. How could anyone expect to learn anything significant about them without a lifetime of effort? I progressed through several failed research ideas involving mangabeys, rhesus macaques, and hamadryas baboons. Slowly, I realized that what I saw as limitations to the existing research on chimpanzees actually presented important opportunities to do better.

As my PhD adviser, Richard provided a strong model for how to build an incredible body of knowledge (and a tremendously successful career) on the accumulation of slow and careful research that is not always aimed at the splashiest outcomes. More importantly, I discovered that the beauty of chimpanzees as a study species lies not in their similarities to humans (which are many) but in the small ways that they differ from us. Finally, I learned to embrace the complexity of designing good research studies on chimpanzees. All those things that make studies more difficult (and results less obvious) turn out to make doing the science more exciting, as you solve puzzles and overcome obstacles one at a time.

A LONG-TERM PERSPECTIVE

For over twenty years, I have had the privilege of working with a team of researchers at the Kibale Chimpanzee Project, which has collected data continuously since Richard began the study in 1987. The scientific value of long-term studies is profound because great apes live very long lives. Everything happens slowly. They grow up slowly, they breed slowly, and their behavior and health are influenced by many years of experience. This means that most studies can capture only a snapshot of the lives of individuals.

Long-term field studies are more than just a string of independent research projects or an accumulation of data. Maintaining these programs is a monumental challenge requiring coordinated efforts at fundraising, logistics, personnel and data management, and engagement with local communities. To even get to the stage of being able to take routine behavioral data on habituated individuals takes many years and huge investments of resources.

While an obvious advantage of long-term studies is the ability to look back at the years of data that have already been collected, an underappreciated advantage is how well these studies enable new research. Instead of starting from the ground up, researchers can implement studies on habituated animals in a well-described habitat, assisted by teams of observers who are intimately familiar with each animal's history. They have the advantage of knowing which chimpanzees are related to each other, approximately how old they are, what their dominance relationships are, etc. Leveraging the strengths of the long-term records, new researchers are able to test premises before going into the field and can employ more creative and ambitious research protocols. These long-term efforts also build research communities that can have tangible impacts on local communities in ways that profoundly influence conservation.

My career has benefited greatly from my involvement with long-term research studies, particularly with the Kibale Chimpanzee Project, and I am proud to have brought my own expertise into building the core research program. I would not be able to do the kinds of research that I do without the efforts that dozens of individuals have put into the project. Given the benefits that long-term field programs contribute to the field and to student training, as a research community we should work to balance our use of these resources with investments in the intellectual, technological, and monetary stability of major field sites for the future.

EMOTIONALLY EXPRESSIVE ANIMALS

Chimpanzees are a remarkable study in contrasts. They can be brutal and unfair. For example, male chimpanzees routinely attack females, sometimes beating them for a minute or more, even using branches as weapons. These attacks seem to come out of the blue but are part of a persistent campaign. The same males that do this can, only hours later, play delicately with an infant. It is shocking and fascinating to experience how quickly tensions turn and how complicated social relationships can be. While some evolutionary scholars have emphasized what chimpanzees can tell us about human peacemaking, others have pointed to chimpanzees as models for the evolution of human violence. The amazing thing about chimpanzee societies is that both of these are true simultaneously.

Chimpanzees are clearly very "in touch" with their emotions. Having experienced this, I think that humans who strive for this have no idea what they're getting into! Often this is most apparent in expressions of fear or excitement. Any disruption to the peace results in a noisy eruption of hoots, barks, and screams. Males exploit this to their benefit by deliberately disrupting others with noisy displays, which can involve crashing through the undergrowth, dragging branches, flinging objects, and drumming their hands and feet against the buttress of a tree, sending all chimpanzees in their wake fleeing. I have seen chimpanzees react with abject panic to harmless things, like bush piglets, and later walk calmly around a viper. When some chimpanzees are truly distressed, which can come from a threat real or imagined, they throw an all-out temper tantrum, flailing and screaming like a toddler, or in more serious cases, scream until they are hoarse and choking.

Expressions of pleasure are quieter but just as emotionally intense. When eating foods they really like, chimpanzees make noises that sound something like a cross between a chirp and a grunt, and it seems as if they simply cannot suppress their pleasure (though in fact they can). A chimpanzee laughing is the most delightful sound. It is the least raucous noise in the vocal repertoire, a gentle staccato panting that accompanies play. When you hear it, it highlights moments of rest and harmony.

NAMBI'S RELUCTANT TOLERANCE

Researchers have faced two impediments when studying female chimpanzees. The first is that females are less gregarious than males. Male chimpanzees within a community like to hang out in large groups, both to keep their brothers and allies nearby and to keep tabs on their competitors. Females, on the other hand, prefer to avoid the feeding competition that comes with large groups. Even when in groups, females are usually quieter and less conspicuous than males. The second is that females are less readily habituated to human observers than males. Since females disperse between communities at maturity, even communities that have been studied for decades will have new females that are not yet comfortable with human observers. I had done my homework. I anticipated these problems. What I had not anticipated is that one of the chimpanzees might just plain not like me.

Nambi was the first chimpanzee I encountered in the wild. When we first met, she was sitting on a low branch of a tree chewing a wadge of fibrous plant material, looking completely nonplussed by the presence of a new human. Nambi was the alpha female in the Sonso community of chimpanzees in the Budongo Forest of Uganda. She was at least in her thirties, and most likely well into her forties, and was the mother of two young offspring, who were playing with branches nearby, and at least two fully grown males. She had been around the block.

After a few minutes, Nambi slowly descended the tree, and with a lingering skeptical glance back at us, meandered down the trail. I was not ready to be done with her. My goal for my dissertation research was to follow females across their reproductive cycles and collect urine and feces for hormone analysis through whatever noninvasive means possible. Others had collected biological samples from wild primates with varying degrees of success. But I wanted *a lot* of samples—enough to be able to track the hormonal ups and downs that occurred during female menstrual cycles and link them to diet and behavior. I had been warned that this would be hard—so hard, in fact, that many of my grant reviewers suggested that it was not worth trying. I was determined. I shot

Nambi

my field assistant a worried glance, and we took off after Nambi. We followed her for a few "blocks" in the forest. She turned a corner and suddenly was gone. We searched all over, stopping periodically to listen for rustling or the small noises that young chimpanzees cannot help but make. Nothing. As big as they are, chimpanzees can all but disappear in a dense forest. Giving up, we headed back to our original location to see if other chimpanzees had come to feed. Nearly there, we heard a small sneeze. There was Nambi, sitting quietly about ten meters off the trail in a dark clump of undergrowth just where we had lost her. Sitting, waiting, watching until her infant gave her away. Nambi glared at us for a minute and then slowly made her way back to where we had originally found her.

We easily found Nambi most days, on her own or in groups. She was clearly unafraid of human observers, often emerging from the forest quietly near one of the field assistants. But I began to notice that we had collected far fewer samples from her than from the other chimpanzees. Perhaps, I thought, this was because she received so much harassment from males. When Nambi's sexual swellings were at their peak, there would be frequent eruptions of male excitement, including displays and threats in her direction. She would urinate and defecate from fear while she fled screaming. These samples were near impossible to collect. Another common situation had Nambi sitting up in a tree with several males crowded around the base of the trunk waiting impatiently for her to come down and follow them. Duane, the alpha male, was particularly enamored of her. He would sit awkwardly staring up at her, periodically shaking branches, stamping his feet, rocking back and forth, and flicking his penis to get her attention. She was remarkably good at ignoring these suggestions. Eventually, he would lose patience and run up to get her. As she avoided him, she would urinate, right on top of the other males, where I could not go.

We began to really focus on getting samples from Nambi. Yet, every time we followed her, she would lose us. It was almost comical. She would walk slowly and confidently until reaching a comfortable distance and then dart around a corner. She would make complicated stream crossings and wait patiently on the other side until we crossed, and then she would disappear or cross back. Because the undergrowth in the forest tended to trap a lot of the chimpanzee urine as it fell, I brought a tool to help reach up into the vegetation to collect it. It was a camera monopod, extendable to about 1.5 meters, with a triangular stand on the end that was perfect for holding a small plastic bag to catch urine drops as they fell. We made sure to use the pole only when the chimpanzees were high in the trees, and most of them simply ignored it. Nambi did not react to the pole, but upon hearing her urine hit the plastic, she barked and moved away. Over time, she began to strategically readjust her position to avoid the plastic bag, stopping midstream to move a few feet away. Her daughter, Nora, meanwhile, would happily run over and aim carefully to poop on our heads.

One day, we found Nambi behaving very nervously. At first, I assumed it was about us, as usual. We soon realized that her anxiousness was directed at Duane, the alpha male, who was nearby staring intensely at her. He was slowly enticing her away from the group. He would move a few steps ahead and wait, scratching nervously. Then he would flick the vegetation and slap his hand on the ground. Eventually, he would approach her as she pant-grunted in submission and embrace her or softly bite her on the back. She would reluctantly follow for a few meters and then stop, and they would repeat their negotiations. In contrast to her past behavior, she seemed almost eager for us to follow.

Duane was visibly frustrated. This happened over and over again, while she looked back at us and in the direction of the other chimpanzees with increasing urgency. Duane was attempting to lure her into a one-on-one mating consortship, something that observers rarely witness. Some have described consortships like "honeymoons," where the two could escape the chaos of the group. They are anything but. Female chimpanzees aren't into monogamy. They mate promiscuously, convincing each male that he may be the father of her future offspring, while managing the attempts by males to harass, intimidate, and guard them. Nambi faced an immediate threat from Duane if she refused his solicitations, but if she drew attention to herself, she might face retribution from other males. Watching this unfold over a couple of hours was agonizing. Duane showed increasing agitation, while Nambi began to show expressions of fear. It became clear to us that Nambi would eventually refuse to follow and that Duane's response would be swift and violent. Finally, it happened. She stared at us for several seconds, turned to Duane, then back at us. I felt her look straight into my eyes. Then, with absolute resolve, she stood up and did something she had never done before—she walked straight toward us and away from Duane. We realized what was happening at the same time Duane did. We cursed. Duane blew up. His hair immediately stood on end, and he began to charge fiercely after her. Her gambit paid off. She lost her nerve at the last second, but so did he. A few body lengths away, he veered off in one direction to avoid us, while she ran screaming in the other. We managed not to soil our pants.

The stories of Jane Goodall and Dian Fossey emphasize the special connections they made with the apes, and it is difficult to avoid the urge to make that kind of connection. And while all ape researchers now take great care to maintain a safe distance from subjects and avoid distracting them from their natural behaviors, it is still hard to maintain an emotional detachment. As the chimpanzees' lives unfold around us, we identify favorites and sometimes imagine that they, too, have favorite humans—that they would not just tolerate our presence but like to have us around.

I have had personal favorites, those who exhibited a certain amount of charisma or who seemed particularly adept (or in some cases, particularly inept) at managing political relationships. Nambi was never my favorite nor I hers. But she challenged me, both in terms of implementing my research and in understanding the obstacles she faced in her everyday life. The limits of her tolerance of being followed around by annoying researchers reflect some of the constraints that female chimpanzees face in associating with other chimpanzees. She could choose to opt out. Yet she showed us very clearly that even as she strove to limit her exposure to us, she had a specific understanding of our role in her world and how we might be manipulated to her favor. Since this early experience, I have been fascinated with the social world of chimpanzees. Through Nambi and other females like her, I have learned that females cannot be dismissed as merely asocial or socially unsophisticated. They seem to value their time away from others and are more discerning than males about who they associate with, when and why. I see some of myself in this. This understanding continues to motivate me to understand more about the inner lives of female chimpanzees and how they construct their social worlds.

WHERE WILL THE RESEARCH TAKE US NEXT?

While chimpanzees are one of the best studied of all primate species, our years of accumulated knowledge continue to generate new questions and changes in perspective. Early work focused largely on describing behavioral and biological characteristics of the species. Later, the focus began to shift to explaining behavioral diversity across populations in light of genetic, ecological, and cultural processes. Researchers were then able to use chimpanzees as a model species to perform robust tests of evolutionary hypotheses about reproductive strategies, social relationships, and collective behavior, among other things. New statistical approaches and methods for applying genetic, hormonal, nutritional, and other types of laboratory analysis to field settings have subsequently allowed us to refine our understanding. While all of these remain ongoing objectives in the field, where do we go next?

Now that several long-term research projects have collected decades of data on individuals, I am excited to study life course development. That is, instead of using simple predictors like age or sex or dominance status, we can begin asking how accumulated experiences shape individual strategies, relationships, temperaments, and health and fitness outcomes over time. For example, while it is plausible that early developmental stress imposes costs that limit survivorship and fertility, can individuals compensate for this by adopting different behavioral strategies or through physiological adjustments? We can also begin to gain a better understanding of interesting behavioral variation in chimpanzees. For example, given that a male's dominance rank is an important predictor of their reproductive success, what factors determine which males get to be high ranking? And is this status costly in the long run?

Unlike for humans, we can gather incredible amounts of objective observation of chimpanzee life experience. At the long-term research sites, many chimpanzees, some now in old age, have been directly observed for one-third or more of their waking hours, with rich details on everything they've eaten, all of their fights and their sexual conquests, the development and breakdowns of friendships, and changes in the environment around them. Some individuals have contributed thousands of samples for assessment of stress, energetic condition, reproductive function, and pathogen load. Every year these datasets grow. No human dataset looks anything like this.

These questions are incredibly complex and will challenge us to come up with new ways to analyze our data. Increasingly, these questions will depend on and will cement collaborative relationships between field projects and will blur the lines between technical specialties.

Chimpanzee research has had a long history of illuminating the evolutionary history of humans. Rather than exhausting the questions, this history has paved the way to addressing new, even more exciting areas of research.

11

DAVID KONI

David Koni is a field assistant working for the Goualougo Triangle Ape Project, a conservation and research initiative based in Nouabalé-Ndoki National Park in the Republic of the Congo, near the borders of Cameroon and the Central African Republic. There, chimpanzees and western lowland gorillas live sympatrically. He has worked with the project's co-directors, Dr. David Morgan and Dr. Crickette Sanz, for over ten years, and his knowledge of the local botany helps the research team better understand the primates' feeding ecology.

GRÉGOIRE, GASTON, AND THE GOUALOUGO

My country is home to thousands of wild chimpanzees, mostly in the very north, where I now live and work with the Goualougo Triangle Ape Project. Perhaps some would imagine that my first experience with chimpanzees would be with those in the forests of my native country of the Republic of the Congo. But this is not true.

When I was nine years old, I saw my first chimpanzee at the Brazzaville city zoo. His name was Grégoire, and he was an adult male chimpanzee who was quite famous in the city. The zoo was free, and many people visited Grégoire every day. I remember visiting and becoming very curious about apes living in captivity, especially since my country was home to both chimpanzees and gorillas in the wild. But Grégoire lived by himself, as did a young male gorilla that also lived there at the zoo. As I left the zoo that day and reflected back on my encounter with these two apes, I felt that something was not right about the way they were living. I could see that their living conditions were poor, and they were not happy. Now, many years later, I can fully appreciate how very different these conditions were from their species' natural environment. But shortly after that encounter, I embarked on a journey that would bring me closer to chimpanzees and gorillas that lived freely in the forests of Congo.

It began with a tradition in my family: each vacation, my father sent his kids off to visit other family members. This was quite a commitment, as I was one of six children. I was particularly interested in visiting an uncle of whom my father frequently spoke at great length. His name was Koni Ngobolo Gaston Nick. Gaston was known to have a wonderful sense of humor and he loved children, and my father joked that he was worried I would end up wanting to spend too much time with my uncle if I visited him! In the summer of 1999, it was finally my turn for a vacation, and I decided to visit my uncle Gaston. I traveled to the far north of the country, to the Sangha, which is a region dominated by a large, continuous forest. I was born at the southern tip of the Sangha region, in the town of Ouésso, before moving to Brazzaville in the south, where I grew up.

My uncle was the chief of Bon Coin, which means "Cozy Corner." It was a small village located on the east bank of the Sangha River where three countries came together at one point. Across the river was Cameroon, and just to the north was the Central African Republic. As there were no national roads traversing the forest interior, the river was the main transportation route. There were few people in the region, and they all seemed to live in villages along the Sangha River. I remember arriving in Bon Coin and thinking I had never stayed in such a small village! There were only around 110 people, most of whom were indigenous peoples living in houses constructed of vegetation and they cooked their meals outside with firewood gathered from the surrounding forest. Plants, many that I had previously seen only in books or on rare occasions on the television, were seemingly everywhere, and the diversity sparked my interest in flora that continues today.

I can remember the first night in Bon Coin. Shortly after falling asleep, I heard something rustling in the small garden behind the house. My uncle awakened to find me looking out a window. We turned on a flashlight and were delighted to see a brush-tailed porcupine digging in the garden soil. This was just the beginning of my animal observations during that visit. The next evening a male forest elephant slowly meandered through the village. Apparently, he was becoming a regular visitor.

After taking a couple days to get accustomed to village life, I accompanied my uncle on foot down toward Bomassa, which was also the site of the Nouabalé-Ndoki National Park headquarters. Compared to Bon Coin, it was a hive of activity. I watched as scientists discussed their missions, assembled their teams, and packed camping materials. The national park was only seven years old at that time, and I remember feeling the buzz of energy from the workers busily moving from one place to the next. My uncle was proud to be among the first people employed by the Wildlife Conservation Society, which was responsible for managing the protected area. Gaston knew the large, winding Sangha River like the back of his hand, and he was employed as the park's first boat driver, regularly making trips to and from towns like Ouésso to transport materials, food, and researchers to the park.

One day my uncle and I were returning late in the afternoon to Bon Coin when we caught a fleeting glimpse of a wild chimpanzee in the low canopy of a tree next to our path. Upon seeing us, the chimpanzee quickly disappeared into the dense forest. I was amazed and so excited to see my first wild chimpanzee, and my uncle was so happy to see my response! Over the course of my visit, I continued to accompany him to work, and on another morning we came across a young gorilla feeding on fruit. Like the chimpanzee, the gorilla also fled immediately by hugging the smooth tree trunk and sliding down and then dashing into the forest. My uncle took me over to inspect the fallen fruits that the gorilla had discarded. I had never seen this type of tree before, and I asked my uncle what the name of it was. He called it a parisolier, or *Musanga*, tree. The leaves were unmistakable—long, large, and grouped with twelve other leaves on a single branch. The combination of the branch and leaves looked like an umbrella, hence the name parisolier, derived from the word parasol. I knew my first tree!

As my two-month summer vacation came to an end, I felt my life had profoundly changed. I had become passionate about observing wildlife and plants, and I knew I would return to this very special forest someday. When I returned home, I spoke with my family of my experiences at Bon Coin and my desire to return and live with my uncle someday (just as my father had predicted!).

Years passed before I traveled again to the north of the country, close to the Ndoki forests. This time, I was off to Pokola, a rapidly growing logging town where two of my sisters lived. It was in Pokola where I met another influential person in my life: Grégoire Kosa, my older sister's husband. Kosa had recently started working in the forestry department for the large industrial logging company, Congolaise Industrielle des Bois, which was headquartered in Pokola. While taking a walk with Kosa one day, we began talking about our shared interest in plants. I was astonished at his vast knowledge of trees. I asked him if we could take daily walks to look at the trees together and to learn more about the plant world all around me.

When my uncle Gaston visited later that year, I expressed my desire to return to the Ndoki forests with him. He agreed to speak with my father about this sensitive subject, and, to my great delight, my father agreed to support my move there. It was at this time that my uncle spoke of an American couple, Crickette Sanz and David Morgan, who were studying chimpanzees in an area inside the park known as the Goualougo Triangle. He mentioned that they were passionate about apes and the forests and committed to helping the Congolese become scientists and conservationists. I was anxious to meet them to see if there was any way I could get a job working with them to help protect the amazing forests that I loved so much.

When I was first introduced to Crickette and David, they both spoke of their great respect for my uncle, his work at the park, and his positive influence in Bon Coin. They needed help in expanding their projects, and, to my excitement, they agreed to take me on! They proposed I start by working on a project counting how many chimpanzees and gorillas were in the area. We did this not by finding the apes themselves but by counting the apes' abandoned sleeping nests. I was an eager employee, interested in learning about the trees and wildlife, but at first I found the work incredibly tough. We would make our way through dense forest thickets in a straight line following a compass bearing, carrying all our camping materials on our backs. All the while, we carefully looked for ape nests, which could be anywhere—on the ground or all the way up in the tops of the tallest trees! As the days turned to weeks, however, I adapted and became more and more encouraged by and engaged in the work. My teammates were very supportive, and I was learning how to identify the differences between chimpanzee and gorilla nests and about the trees and ground vegetation or herbs the chimpanzees and gorillas frequently used to make their nests. Each species had preferences in nesting materials, and I found the nests could be very elaborate with combinations of foliage carefully bent together to make a comfortable resting site.

The most exciting part of the work, however, was when we came across the chimpanzees and gorillas themselves. On several occasions, in distant forests, chimpanzees responded to our presence as if they had never seen people before. Rather than running away, the apes surrounded us and came closer as if to get a better look! The research assistant in charge of the project was Crepin Eyana Ayina, who had been working with Crickette and David for five years. He recognized my interest in observing the chimpanzees and thought I would really enjoy working at the Goualougo base camp, where the chimpanzees were very used to seeing humans and allowed scientists to observe them in their daily lives.

By early 2009, I had two successful transect missions under my belt and I was ready to start working as part of the chimpanzee observation teams out in Goualougo. Like at the start of the transect missions, there was a lot to learn. How to find, follow, and act around chimpanzees were the first steps. I remember being initially surprised by the attention given to having as little influence as possible on the chimpanzees' behavior. My supervisors emphasized how we wanted the chimpanzees to ignore us and go about their daily activities as if we weren't there. Each individual of the Moto community of chimpanzees was easily recognizable based on how they looked and who they tended to spend time with. I quickly grew to appreciate seeing natural chimpanzee behaviors, and I also enjoyed recording my observations and reviewing the data at night, discussing the different individuals and behaviors we had observed in the field.

Along with collecting information on the apes, we also recorded the plants that the chimpanzees ate. Apart from the most common tree species, however, it was clear that much work was needed to properly identify the flora. There was no botanical guide for the park nor an herbarium at the park headquarters where unidentified plant samples could be compared with known specimens. Crickette and David wanted to remedy this, and they had recently started collaborating with David Harris of the Edinburgh Royal Botanic Garden in Scotland. David was a plant taxonomist with expertise in the trees of central Africa. He had spent over twenty years in the region inventorying the trees and had worked intensively with Sydney Thony Ndolo Ebika, another research assistant and an aspiring botanist and educator. It was under David's supervision that Sydney had just received his master's degree in the biodiversity and taxonomy of plants from the University of Edinburgh. I now had two botanists eager to share their knowledge with me. I felt very fortunate to be associated with a project that was committed to botanical studies and to facilitating the mentorship of Congolese researchers in the study of plants. Sydney became my mentor, and I was awarded my first contract as a research apprentice. I not only had a permanent position with the Goualougo project, but my job reflected those specific interests in ape behavior and botany.

DOROTHY THE HONEY POUNDER

So many individual chimpanzees in the Moto community have taught me a lot about chimpanzee behavior, sociality, and ecology. I feel like I learn something new every day when I am out in the forest. However, one adult female, Dorothy, stands out for being particularly influential on me. She was a young adult who had come into the Moto community just before I started working with the Goualougo project. She was quiet and submissive to the older, more established females in the community so it was not uncommon to find Dorothy off on her own or in a small group. When she gave birth to her first offspring, a male named Oz, she was learning how to be a mother while still trying to fit in with the other chimpanzee community members.

What I learned most from Dorothy was how chimpanzees use tools to gain access to hidden food items they otherwise would not be able to get to. In the Goualougo Triangle, chimpanzees have been seen to use multiple tools in different contexts, such as termite fishing, ant dipping, and something called honey pounding. I was astonished the first time I observed Dorothy perform this last behavior. On that day, Dorothy and Oz were feeding on fruit in the canopy when suddenly Dorothy stopped. She traveled along a large branch high up in the treetops and crossed into the canopy of a neighboring tree with Oz trailing closely behind. I tracked her movement from far below, using binoculars as she was over forty meters up in the trees. Dorothy was traveling as if she had a purpose and only stopped for a moment to break off a medium-sized branch, which made a loud popping sound when it snapped. Initially, I thought she was going to make a day nest, which would not have been uncommon. But I was wrong. I watched as she carried the branch to where a hive of stingless bees was located.

Dorothy positioned herself just to the side of the hive entrance, placing her free left arm around the large branch, almost as if trying to hug it. Without hesitation, she quickly drew her pounding club back then plunged it forward with impressive force, pounding the hive entrance with the blunt end of the tool. Her pounding precision was just as impressive as her apparent determination to break into the hive. Thousands of tiny, stingless bees began swarming around her. After several minutes of pounding, Dorothy moved off a short distance but soon returned with a much smaller twig. She jammed this new tool into the hive entrance, moving it around inside. Upon withdrawing the tool, she inspected and licked the end, which was coated in honey. She repeatedly inserted and extracted the tool, greatly enjoying the sweet rewards of her

Dorothy

efforts. It had been no easy task: so high up in the canopy, she had risked her life and used multiple tools to get at this prized resource. And I was very impressed when Dorothy kindly responded to Oz's begging by sharing the sweet honey with her young son.

LIFE LESSONS FROM THE CHIMPANZEES

I think my work with chimpanzees has greatly influenced me as a person in several ways, particularly when it comes to my awareness of the social interactions in my life. To see how social the chimpanzees are and how they invest in relationships with each other is a good reminder of how to be a good human being.

Early in my career, I started thinking about how important it is to invest in family and about what kind of father I would want to be. I want my wife to have all the support she needs in taking care of our children and for my children to feel like they can pursue their dreams as I did. When I am working out in the Goualougo Triangle, I am mostly out of contact with my family, as there is no cell coverage or internet access, except at base camp. As a result, I value the time I have with my family more than ever and try to never take our time together for granted. None of my siblings live in the same town as my family; this is so different from when I grew up. As a result, I spend a great deal of time speaking by phone with my sisters and brothers to stay in touch.

Of course, choosing a career studying wild chimpanzees has come with challenges as well. Early on, my father spoke of his expectation that I would pursue a career in the military. He served in the military for over thirty-five years and was proud of his promotion to an elite group called the "Guard of Commandants." In fact, I had several older brothers and sisters who had already begun their military careers as I was growing up. By the time I was nine, and had met Grégoire the chimpanzee, I had an older sister who was already a lieutenant in the military, several siblings who were captains, and my oldest brother was on his way to becoming a colonel. My father considered these to be good-paying, stable jobs with opportunities to advance. However, I did not like the idea of being a part of the military institution. The idea of carrying a weapon and potentially having to use force in violent situations was not of interest to me. Even from a young age I knew I wanted to be outdoors, and then after my visit to Bon Coin, I had more of a sense of direction. I wanted to work in the forest with wildlife.

It was extremely difficult to convey to my father my wish to study wildlife and plants. I admired his career and his work ethic, and I did not want to let him down. I was thankful for his concern about his children's futures and his desire to help us the way he felt he best could. Over the years, however, my father began to accept my choices. He came to understand my passion for working with wildlife and studying forest ecology, and he eventually accepted my career path. I am so grateful for that support and his willingness to acknowledge my own interests.

Early on, my father and family asked more about the work, and the details of my day-to-day activities, than about the chimpanzees. But I took time to share details about the individual chimpanzees I had been observing, their behaviors and personalities. More and more, my family became interested in hearing these stories about my latest observations, the chimpanzees' tool use, and how young chimpanzees learn from their mothers. Over time, I think my work in the forests has begun to change how they see chimpanzees and the importance of wildlife in our country.

Reflecting on my work with chimpanzees, I am most proud of the fact that my father came to greatly appreciate my career and my enthusiasm for chimpanzees and plants. When I return from the forest, I always make an effort to call him. He loves hearing the stories from the forest and especially about the lives of the individual apes. He also started requesting a bottle of water from the Ndoki River, a medium-sized river that forms the southeastern border of the park that we cross going in and out of the protected area. He calls this fresh water "the taste of the forest." News of this gesture spread quickly in my family and my uncle also began regularly requesting a bottle as well!

THE COURAGE TO FIGHT FOR THE FUTURE FOR CHIMPANZEES IN THE REPUBLIC OF THE CONGO

The greatest challenge facing chimpanzees in the northern Republic of the Congo is that they are being hunted. Historically, there has been a taboo against eating chimpanzee meat in this region, but since I started working at Goualougo, I've noticed how the populations in Bomassa, Ouésso, and Pokola have changed, so now there is real concern. At one time, there was a common perspective and traditions but, more recently, people from distant areas, with different cultures and differing opinions about the forests and wildlife, have become integrated into the population. Not only are they likely to think differently, but there are simply more people in the area now. More food is required, and there is not enough food in the markets. Roads have also expanded into remote regions around the park, providing hunters new access to forests and wildlife. Poachers take whatever they can, without species preference. As more and more people move into the region, the remote forests and chimpanzees in this landscape will increasingly be at risk because of the developing bushmeat networks.

Despite these increasing risks to the chimpanzees and other wildlife, I think it is important to maintain a positive outlook and consider what can be conveyed to the next generation, which will perhaps go on studying the chimpanzees as I have. I try to convey to them how important it is to have patience when studying chimpanzees: get to know the chimpanzees and their personalities, patiently learn how to follow their tracks, and learn the food items they eat during different times of year. I cannot overstate the importance of spending as much time as possible in the forests themselves. Waiting patiently through rainstorms in the forests can lead to incredibly rewarding observations afterward. It can also take time to develop good observation skills and getting to know chimpanzees and different aspects of their lives can make one a better scientist. Working in different forests or at different sites can also be informative. I am fortunate to work at Goualougo and also at another site, Mondika. Our botanical work shows that even though these sites are only sixty kilometers apart, their forests are quite different.

The work can be difficult and tiring, particularly when studying chimpanzees in areas where they are exposed to poaching pressure and other anthropogenic disturbances, such as logging. These troubles can have significant impacts on our work, but remembering how my work helps protect chimpanzees can help tremendously in keeping focused during tough periods. Ultimately, what we need is simply courage. We encounter elephants and gorillas multiple times a day, which can be quite dangerous. After any tense encounter it takes confidence to continue on with following the chimpanzees and to collect the precious data.

But one needs courage not just when out in the forests but also when taking on ideas or initiatives that at first seem bigger than oneself. Years ago, the act of writing a book or telling such a story seemed like something a world away from me, something that others do when they have something to say. At that time, I certainly did not feel like I had a story to tell—or even a voice to tell it. My uncle started writing a book over twenty years ago, and at the time

I did not appreciate the importance of him telling his story. Unfortunately, he passed away before finishing his book. But I like to think that both his story and our family history continue on with me in the present and with my experiences in the Goualougo Triangle. To me, it's also now a much larger story that includes the amazing diversity of plants and the tool-using chimpanzees and the people I have met and worked with to study and protect them all. I now find power in the courage that I did not know existed inside of me and know that I can use that courage to help the plants and animals of the Goualougo Triangle that I have come to love.

12

TATYANA HUMLE

Dr. Tatyana Humle is a professor of conservation and primate behavior at the University of Kent but spends considerable time continuing her fieldwork on chimpanzees in West Africa. Her work focuses on understanding how great apes manage life in landscapes that are shaped by human development, and she strives to improve the coexistence between these two species. To do so, some of her work has been based at the Chimpanzee Conservation Center in Guinea, which is home to approximately sixty chimpanzees, victims of the illegal pet trade.

FROM FRENCH FORESTS TO WEST AFRICA

When I think about it carefully, three critical "spices" combined to brew my fascination with chimpanzees.

The first is the constant lure of the natural world that has been present ever since I can remember. Growing up in rural France, spending more time outdoors than indoors, watching, observing, smelling, questioning, and feeling the natural world around me were all pivotal in cementing my incessant wonder for nature. As a child, I spent hours flat on my belly peering into the underwater world of a small pond in our garden, appreciating the cycles and interactions of life, whether admiring tadpoles or dragonfly larvae, and trying to imagine how it is possible for these living beings to transform so dramatically into adults (quite something to get your head around as a youngster); watching a lizard or a spider stealthily capturing a fly; uncovering mysterious mushrooms in the forest and discovering the intricacies of symbiosis, parasitism, and mutualism that surround us and shape nature; or simply enjoying the rewards of a natural harvest of mushrooms that I had foraged. Of course, these puzzles were key drivers of my budding scientific curiosity, but some larger questions kept buzzing in my mind: How have we evolved as a species to become one that can ponder all these wonders of life and question the whys and hows? How have we evolved to become the main threat to all this natural beauty that surrounds us? And what really makes us different from other animals? The diversity of cultures has also always been a source of marvel to me. Where does it all come from, and what has shaped, and continues to shape, our evolution and existence?

The second spice of inspiration was a National Geographic documentary I watched when I was eleven. It was on the career of Dian Fossey and her life with mountain gorillas. The lush evergreen mountain forest, the gorillas, the people—and among it all, this extraordinarily dedicated woman who lost her life battling for the survival of her nonhuman friends amid an unstable region of Africa. I was inspired and the seed of my own future was sown: if I am to better understand humans, I absolutely must study apes. I set my sights on Africa; my father had traveled along the entire western coast in the 1950s.

The third, and most elemental, spice was Jane Goodall's seminal work, *The Chimpanzees of Gombe*. It was the first English-language book that I ever read from cover to cover. It was a gift from my mother as we strolled through the natural history section of a bookshop in London on my twelfth birthday. It still remains, to this day, the most complete work on chimpanzees in their natural habitat.

With this vivid curiosity for the natural world, combined with the inspiration of these two dedicated women, I was on my way. Before I knew it, it was 1995, and thanks to Professor Tetsuro Matsuzawa, at the Primate Research Institute of Kyoto University in Japan, I was off to Guinea in West Africa. I was then a second-year undergraduate student of zoology at the University of Edinburgh. I soon began learning from the amazing community of chimpanzees living around the village of Bossou, which had been studied since 1976 and were habituated to the presence of human observers. There was no mistaking that my career was set as both my mind and heart were fully committed.

THE TWO VEVES

Every chimpanzee I have encountered over the years, whether wild or captive, has in some way influenced the way I perceive chimpanzees. But two of them, and their stories, profoundly affected me. They share the same name, Veve, which means "sweat bee" in the Manon language.

The first Veve was born in May 2001—the first offspring of a young female named Vuavua from the Bossou community of chimpanzees I was studying in Guinea. At the young age of twelve years old, Vuavua was a good mother, in part thanks to the help provided by Velu, her own mother and one of the oldest females of the community. Sadly, in December 2003, a respiratory epidemic spread throughout the group. It affected all of the chimpanzees: coughing, runny noses, and lethargy were prevalent. During the epidemic, we struggled to locate the chimpanzees; they were unusually quiet, save for the sounds of the sniffling and coughing, and they dispersed deep in the forest. During our furtive forest searches, we encountered the decomposing body of Kai, who, at over fifty years old, had been the oldest member of the community. It was a devastating loss, but it was not only the elderly individuals who were affected by the disease. The next day we found Poni, a ten-year-old male lying prone between the trees. And perhaps most heartbreakingly, we sighted an adult female, Jire, slowly walking through the woods, carrying her deceased infant on her back. Mother chimpanzees will often stay with, and carry, infants long after they die, the maternal bonds desperately persisting.

Clearly all individuals of the group were susceptible to this infectious disease, but we worried most about the infants. When we saw Vuavua without Veve, our hearts sank. Had Vuavua abandoned her sick infant? Had Veve also succumbed to the epidemic? Nine days later, our despair was lifted by news from the village: someone had sighted a baby chimpanzee in a coffee field located between the two main hills most frequented by the chimpanzees. My mind raced: Could this be Veve? Could she have possibly survived without her mother for nine days? We quickly went to check out the location and, indeed, there sat Veve at the edge of the field. She was familiar with us so she did not flee when we approached her slowly. She was clearly weak but did not show evident signs of coughing. She was only two and a half years old—too young to survive on her own, as infants are not weaned until about five years of age.

But, unfortunately, the rest of the group were spending all their time on a faraway hill consuming abundant figs that had recently ripened; they were not likely to pass through the coffee field anytime soon. We had to think quickly. Should we capture Veve and take her to her mother? Should we wait for the group to come this way, but how long would that take? Veve appeared weak, but she was still alert and able to climb trees on her own, so we decided to watch over her from a distance to avoid causing her any unnecessary distress. We provided her with fresh clean cultivated fruit that she would be familiar with, such as pineapple, banana, oranges, and papaya, as well as the fruit and pith of terrestrial herbaceous vegetation. We watched with amazement that evening as she actually built her own nest high up in a tree. It was remarkable to see as it is quite rare for infants of that age to construct a suitable nest and sleep on their own. Miraculously, on the third day, the chimpanzee group headed into the coffee field. Vuavua bounded over upon seeing Veve and immediately scooped her up. Back in the loving arms of her mother, Veve was surrounded by virtually all the chimpanzees in the group and was being thoroughly coddled and groomed. They were clearly all excited by her retrieval, and of course we were, too!

Over the coming days, Veve appeared to regain her strength, but we were still concerned. She mostly fed independently on solid foods and foliage but only suckled for very short durations. This was concerning as infants of that age typically receive most of their nutrition from their mother's milk. We presumed that she had become more autonomous during the time away from her mother, but what we did not realize at the time was that her mother's milk had dried up. Veve would try to nurse, but, unfortunately, she was not getting enough nutrition. Two weeks later, Veve passed away.

Vuavua carried Veve's dead body for nearly three weeks, evidence of the strong bond between them. My hopes, fueled by Veve's resilience and amazing ability to survive on her own at such a young age, were crushed. Her passing was disheartening and made me also realize how susceptible chimpanzees can be to disease, especially human-borne diseases. Although we were never able to identify the pathogen that was at the origin of this epidemic, we are suspicious it was transmitted by some tourists who came to see the chimpanzees some days earlier and who had failed to declare that they were ill. The epidemic was a particularly sad time for all of us and marked a significant decline in the Bossou chimpanzee population from which they have never recovered. As a result of the outbreak, significant changes in the policies around observing chimpanzees were implemented. Protocols that are now so familiar to us as a result of the global COVID-19 pandemic, such as wearing a mask and maintaining a safe distance from the chimpanzees, were made mandatory. In a way, these protections for all the remaining chimpanzees are a legacy of young Veve.

A few weeks after her passing, while at Bossou, I was approached by the director of a nearby chimpanzee sanctuary—the Chimpanzee Conservation Center—to assist her with a confiscation. A two-year-old chimpanzee was being kept as a pet by a Guinean family in the nearby town of N'Zérékoré and someone was needed to take the chimpanzee to the sanctuary. When I consulted with the local authorities in charge of confiscations, they told me it would be difficult to recover the baby chimpanzee from the family as they had hidden her and were refusing to hand her over. Together, we went to the house and sat down in the courtyard to discuss the situation together with the family. They told us how the chimpanzee infant had been the father's gift to his eight-year-old son, who had just lost his mother and baby sister in childbirth some few weeks back. The father was living and working in faraway Mamou at the time and this infant chimpanzee was to be like a little sister to his young son.

The pet trade is a significant issue in the region and Guinea was recently banned from the Convention on International Trade in Endangered Species of Wild Fauna and Flora (CITES) for the illegal trafficking of live wild-caught young chimpanzees for destinations such as China. Hundreds of young chimpanzees have been illegally captured and exported. While many have been confiscated by the local or national authorities and are being rehabilitated at the

Chimpanzee Conservation Center, every single capture of an infant typically implies the death of one or more adult chimpanzees. The pet trade can therefore have a dramatic cumulative impact on chimpanzee populations, not only causing a decline in numbers locally but significantly disrupting their social structure and their behavior toward people.

When speaking with the family in N'Zérékoré, it became clear why there was a strong reluctance to hand the baby chimpanzee over to the authorities. With this in mind, I carefully tried to explain why a chimpanzee needs to live among other chimpanzees, not humans. I told them how this baby chimpanzee had been cruelly orphaned, that her mother had most probably been killed so the baby could be sold, and how, at such a young age, she was unlikely to survive. I noticed everyone in attendance, including the young boy, reflect on my words. I saw a glimpse of sadness and understanding. The family then proceeded to consult one another, and one of the adult cousins then left the courtyard and returned five minutes later with the baby chimpanzee in his arms. She looked relatively healthy, albeit with a swollen stomach (probably parasites), but she had no wounds or injuries.

When the cousin handed the infant chimpanzee to me, she quickly grasped my arms for security. I promptly arranged to buy some water, rehydration salts, and a traditional cloth so I could carry her on my back. After eight years studying wild chimpanzees, this was the first time I ever held a baby chimpanzee in my arms; fortunately, I knew how to behave, but I had never imagined this would ever happen. The little boy said goodbye to her, and I could see they had become attached to one another, but I told him that he could visit her anytime and he had made the right choice for her future. She would be better off with other chimpanzees, good veterinary care, and a more natural environment.

Veve

I decided the infant should be named Veve, after the young and resilient chimpanzee I had tried to help at Bossou. Days later she safely arrived at the sanctuary.

I visited the sanctuary years later and was excited to see how Veve was doing. She was happily integrated into a group and learning what it takes to be a chimpanzee. Veve's story embodies the sad reality of the pet trade in Guinea and elsewhere across Africa but also the amazing resilience of babies and their astounding will to live when the right support and care are given to them. But in spite of her resilience, and even with an excellent and nurturing environment in the sanctuary, Veve has had her freedom stolen from her. Unfortunately, releasing individuals back to the wild is a complex enterprise that is not an ethically suitable solution for all rehabilitated orphans, especially those who have had traumatic experiences when kept as pets.

These two Veves have taught me so much and made me deeply reflect on the threats that this great ape species and many others are facing today: their vulnerability to disease, human ignorance, and the challenges of people and wildlife coexisting. At the time, it became blatantly clear to me that we also need to understand people's motivations, perceptions, and behavior if we are to help conserve chimpanzees. But these two Veves also remind me of how amazingly intelligent chimpanzees are and how adaptable they can be in spite of all odds.

WHAT DOES THE FUTURE HOLD FOR CHIMPANZEES?

For all of us studying chimpanzees, special individuals like Veve inspire us to do more to understand and conserve the species. But to truly understand the threats that chimpanzees face, we must take a broad view of their situation across Africa and how their habitats are inevitably threatened by human activities.

We know that wild chimpanzees have a wide but discontinuous distribution across Equatorial Africa. They occur from southern Senegal, across the forested belt north of the Congo River, to western Uganda and western Tanzania, and in areas ranging from sea level up to 2,800 m in altitude. They inhabit a range of habitats, primarily moist or dry tropical forests, but also more savanna-dominated landscapes interspersed with riverine forest and degraded landscapes dominated by cultivated fields and fallow areas. The species has truly adapted to a variety of habitats. Overall population estimates range anywhere from 172,700 to 299,700 individuals, but this may well be an overestimate as those figures are now over fifteen years old and populations have been declining across the continent. According to the International Union for Conservation of Nature (IUCN) Red List of Threatened Species, chimpanzees in West Africa are now critically endangered, meaning that the populations could decrease by more than 80 percent over the next three generations! This is, unfortunately, a likely scenario as the majority of remaining populations live outside protected areas in landscapes that are increasingly being modified by human activities, infrastructure, and presence.

Over the years, I have witnessed an exponential expansion of towns and villages and a significant spread of agriculture and industrialized activities. Although chimpanzees do have an amazing capacity to adapt to human-modified landscapes, they can only thrive in such areas if tolerated and if they have access to forest or equivalent natural refuges, where the presence of people is limited. Such a balance can be found in traditional agricultural landscapes, where people practice slash and burn agriculture, but landscapes are changing; I have witnessed this firsthand. Intensified farming practices rapidly deplete soil fertility, thus compromising people's livelihoods, and reduce the abundance of native plants that offer food and shelter for wildlife. Many farmers recognize the deteriorating impact of their farming practices on the environment, but they find themselves unable or ill-equipped to manage this trend. In addition to farming, other resource extraction activities are increasingly unsustainable, such as firewood collection and wild animal hunting. Once local subsistence activities, they have now become commercial endeavors in many areas and have significantly modified the landscapes where chimpanzees and people co-occur. What may have previously been termed a harmonious coexistence between people and chimpanzees in many areas is now being seriously tested.

Further exacerbating these problems is the increasing influence of nonlocal and commercial interests. Not only does this growth affect the resource needs of a local community, but it can dramatically modify the demography, cultural makeup, and religious practices, putting at risk local or traditional environmentally friendly practices and taboos that may have helped protect chimpanzees and other wildlife in the area. This is something I have witnessed at Bossou, where most Manon families recognize the chimpanzee as the reincarnation of their ancestors and thus live harmoniously with them for the most part. The influx of nonresident people into the village, however, has resulted in a dilution of these belief systems; people are now less tolerant of chimpanzee incursions into their settlements or cultivated fields. Such cultural shifts, in addition to issues of development, permitting, and land rights, have created immensely challenging contexts for implementing successful conservation efforts and initiatives, especially where local people feel particularly sidelined in any decision-making processes concerning large-scale development and external agencies.

Most farmers across West Africa are subsistence farmers, so damages to subsistence crops run the risk of affecting local people's livelihoods. People are also typically less tolerant of loss or damage to cash crops, such as pineapple, as they may provide people with their only source of monetary income. Retaliation and aggressive crop protection measures against chimpanzees that eat or damage crops, such as the use of slingshots, can also elicit changes in chimpanzee behavior, who may in turn act more aggressively when encountering humans. This feedback loop aggravates the relationship between people and chimpanzees, further fueling fear, and harming the coexistence between people and chimpanzees.

IS THERE HOPE?

Even if the causes of the decline of West African chimpanzees are largely understood, they have certainly not ceased, and they are not easily reversible, especially outside protected areas. Conservationists and chimpanzee researchers are faced with a great challenge that we alone cannot address. Chimpanzee conservation, and that of other wildlife and natural ecosystems, is vital to the well-being of the local, national, regional, and global health and economy. It requires an interdisciplinary and multidisciplinary approach that must involve all local and national stakeholders. However, the challenges remain fierce and crucial questions remain: Can people and chimpanzees coexist and under what conditions? Is chimpanzee conservation compatible with development? Can different values and interests be reconciled to ensure a balance between people's livelihoods, needs and wants, and conservation objectives? I, for one, believe there is hope. But we must act and do so with some careful consideration.

First, we must recognize the key role that chimpanzee sanctuaries play. There are dozens of large chimpanzee sanctuaries located throughout Africa, some of which I have worked with directly. While many people still think of sanctuaries as refuges for orphan chimpanzees, they have played an increasingly important role in local conservation, environmental education, and local law-enforcement implementation.

Second, we must embrace new technologies. I have increasingly been employing such tools in my own work in Sierra Leone to survey and monitor chimpanzees more effectively and identify areas outside protected areas where chimpanzees live. I am now using drones to survey nests or to generate aerial images of areas where people and chimpanzees coexist to better understand how land-use changes influence how chimpanzees navigate and use these increasingly human-dominated landscapes. Camera trapping is also a powerful tool that can help us with identifying individual chimpanzees and survey populations with minimal human presence in the field. These new technologies can help us better understand how chimpanzees adapt but also what features in the landscape most affect their distribution and persistence.

Third, we must recognize the urgent need for better land use and infrastructure development planning to avoid conflict over land rights and impacts on chimpanzee populations. Likewise, we must improve our understanding of the social, ecological, and behavioral changes affecting people-chimpanzee coexistence and conflicts among international, national, and local stakeholders, especially among the people who coexist with chimpanzees every day and who are, ultimately, the local stewards of biodiversity. At all levels, an understanding of the behavioral ecology of chimpanzees and factors that influence species decline also need to be shared with national and local governments, local people, and relevant NGOs. This information should also be adapted to inform education campaigns and school curricula where relevant.

Fourth, it is necessary to promote environmentally friendly revenue-generating activities and agricultural practices, as well as a thoughtful selection of the types of crops being cultivated (i.e., low-conflict crops or ones that are not consumed by chimpanzees and other wildlife). Without considering sustainable alternatives for the local people, we cannot possibly continue to insist they cease other, more harmful activities.

Finally, but most importantly, there is an urgent need for greater consultation and participation of local people in decision-making processes affecting their environment. Nationals of chimpanzee-range countries should be equipped with the appropriate knowledge and skills to foster and implement sound development strategies at all levels that are compatible with the conservation of chimpanzees and other biodiversity and ecosystem services. Higher-education provisions need to be enhanced to strengthen research capacity across all relevant disciplines, whether agriculture, land-use planning or the social or natural sciences.

This list is not exhaustive, and surely there are many different approaches and schemes that can help conserve chimpanzees both outside and in protected areas, but I have sought here to mention the few that I have seen as possibly the most important based on my long experience in West Africa. But the clock is ticking. The liberal, intellectual chimpanzee in Pierre Boulle's novel *Planet of the Apes* reminds us, "Reasonable humans? Bearers of wisdom? Humans inspired by the soul? No, that's not possible." The success of the conservation efforts for chimpanzees, other wildlife, and their ecosystems, which people's lives hinge on, will very much depend on the extent to which this is proven false. With this growing awareness of the complex issues and challenges at play, I wish to remain optimistic, but there is much to be accomplished and we need to take action now and collaborate with others to achieve the right balance.

13

BRIAN HARE

Dr. Brian Hare is a professor of evolutionary anthropology and psychology at Duke University and founder of the Duke Canine Cognition Center. He has studied great-ape cognition in several accredited African sanctuaries, including Tchimpounga Chimpanzee Rehabilitation Center in the Republic of the Congo, Ngamba Island Chimpanzee Sanctuary in Uganda, and Lola ya Bonobo in the Democratic Republic of the Congo, demonstrating their value for noninvasive research and conservation. He has written and edited a number of books, including Survival of the Friendliest, which he coauthored with Vanessa Woods.

THE GOLDEN RECTANGLE

Like many people who ended up working with animals, I watched a lot of wildlife documentaries on television as a child. Yes, I know I should have been outside chasing birds and butterflies under the Georgia sun, but I was a child of the 1980s growing up in Atlanta and television was central to my young, curious mind. I loved to watch just about anything about animals or animal behavior, but I have to admit that there was one sight that would stir my enthusiasm more than others. I remember clearly the excitement that would arise in me when I would see that golden rectangle representing

National Geographic appearing on the screen. When I saw that, I knew I would be in for something special.

Of course, Jane Goodall was featured in many of these National Geographic programs, and like so many other kids, she introduced me to chimpanzees. Watching her push through the forest in search of chimpanzees and perch on a rock face with her binoculars, listening for the distant hoots and screams of the chimpanzee troop, was fascinating and exciting to me. I think even then, the idea that I could spend my life studying chimpanzees, slowly started to creep into my prepubescent brain and it never really went away.

Fast-forward to high school. All of a sudden, I had people asking about what I was going to do with my life and giving me all sorts of advice of what jobs might be best for me. I had maintained an interest in animal behavior, so a lot of friends and family were suggesting that I become a veterinarian or go into forestry, but these ideas really didn't resonate with me much. However, I found a program at my high school where you do internships at local businesses or organizations and figured out I could do a placement with Zoo Atlanta. I will admit the motivation to participate in the program was enhanced by the fact that you could leave school really early, but the opportunity to work around animals was pretty great too. I started on a study collecting behavioral data on the drills, which are monkeys and close relatives of mandrills and baboons. They were amazing to watch, and it did not take long for me to get hooked on primates.

Interning at the zoo gave me a great opportunity to not only see and learn about the primates I was studying but all the other animals as well. The drills lived right next door to the gorilla exhibit, so when I was up on a rooftop watching the drill troop, I could also see the gorillas. There was a lot of excitement those days because Willie B., the famous silverback gorilla at the zoo, was going to be a dad. He was a big deal there because of his amazing story. He was born in Africa and was brought to the zoo in the early 1960s. He was named after Atlanta's mayor at the time, William Berry Harts-field. Zoos were very different back then, and Willie B. lived alone for many decades with just a television to keep him company. But when the zoo opened its brand-new outdoor gorilla exhibit in 1988, it was amazing to see this giant silverback, who had gone so long without proper socialization, come into his own. He sired a bunch of offspring and led that troop of gorillas for many years after. He is a legendary gorilla in and around Atlanta, even many years after he passed away.

Being around all that gorilla excitement only got me more interested in studying primates. I remember Taz, a young male in the group, throwing rocks at me as I walked up the trail to my research station and really feeling amazed at how great it was to be part of that whole environment. Not everyone gets to be harassed by a young ape every day!

Next, it was time for college. My mom knew I was interested in primates so she gave me two books by Frans de Waal. Little did I know that, years later, Frans and I would work together! To be honest, I don't think I actually read those books until much later, but they still inspired me because they reminded me that there were folks, right there in Atlanta, doing the type of work that I was interested in: learning about primate behavior and learning. Even though I was probably thinking more about sports than primatology those days, I did want to go to college and learn more about animal behavior.

I got a scholarship to Georgia Tech, but my parents really encouraged me to go to Emory University because the Yerkes National Primate Research Center was there and they knew it would be a great place for me. I hate to admit it, but they were right. I ended up taking a class with Frans de Waal, whose books I had only skimmed before taking his class, but it was through his lectures that I was introduced to bonobos and really got hooked on the idea of studying them like Frans had at the San Diego Zoo. Things just fell into place from there. I ended up meeting Mike Tomasello and Josep Call and others doing amazing work around great ape cognition and behavior, which led me to follow that career path. It started with that golden rectangle on the television and led to a lifetime's career of working with both chimpanzees and bonobos.

SANCTUARIES, THE NEW FRONTIER

One thing I am very proud of is the role my research team and collaborators have played in demonstrating that accredited sanctuaries are a relevant and valid place in which to do science. Not that long ago, scientists used to scoff at the idea of conducting research in a sanctuary setting, especially those in Africa!

I remember, in 1999, when the director at Yerkes sent out an email saying that all the chimpanzees were going to be moved out to live in accredited zoos and sanctuaries and that the chimpanzee research there was effectively over and everyone needed to prepare for that. Everyone was a bit panicked that the research they had worked on for so long was going to be suddenly cut short because there were no other options. I was told so often that there was simply no other way to do experiments in cognitive science if opportunities no longer existed in lab settings. I remember being told that I would have to choose between being involved in "serious" science or to focus on conservation and welfare topics. Richard Wrangham, my mentor, deserves the credit for pointing out the potential of African sanctuaries as research sites.

Throughout my career, I have had the fortune of working at a number of incredible sanctuaries and with the apes that called them home.

JESSIE AND THE GLOVE

I am very fortunate because I have met a lot of chimpanzees and bonobos. I have worked in laboratories, zoos, and sanctuaries, both in Africa and the United States. And I have even followed around some wild chimpanzees for a short time, but mostly I have studied apes in captive settings where I can give them novel puzzles to solve so I can study how they think. So, thinking about specific individual apes that have had a big influence on me is tough. Not because they haven't had an impact on me but because so many of them have, and in so many different ways.

One of the most striking things about these apes is how much individual variability there is. They all have their own characteristics and their own stories. I remember Frans de Waal saying to me in a freshman class that it is so important to just observe the animals you study and not necessarily interact with them directly. And, of course, this is absolutely correct for the type of research he does, but most of my work is really interactive. Most of my experiments involve giving the apes things to do and problems to solve so we are really engaged with each individual and seeing their personalities come out when they are faced with the different types of challenges we give them. They can take different approaches to the same problem, but they also take different approaches in getting to know me, the guy working with them on these studies.

For instance, a chimpanzee named Phineas was really intent on teasing me and, in a way, testing to see how far he could push the envelope when we first met. He was very interested in determining if he could manipulate me, or if he could dominate me, and he was always testing to see what he could get away with. Then there was another chimpanzee, Tai, maybe the first chimpanzee I ever actually met, who was just so happy to meet me and gave me nothing but love. It was like I was the son she never had. She was obsessed with wanting to groom me through the wire mesh of her enclosure and buttoning and unbuttoning my sleeve cuffs. I really had nothing but positive interactions with her from the start and she seemed much more interested in me than the tasks I gave her to do.

Then there was a chimpanzee named Jessie, a young female who was so sweet and calm and always interested in the things we had her do. She was fascinated by new things, and I remember one day she kept pointing at my hand over and over. I thought she wanted my watch (which I would not have given her), but she eventually made clear she wanted one of the latex gloves we all wore at the lab. She was so gentle but persistent. I couldn't resist. I never should have done it, but we had this really great connection and I thought, "Of course I can trust her—she can look at the glove, and maybe play with it a bit, and then she'll hand it back to me." So I stripped off the glove and handed it to her. She took it gently from me, and I was ready to have a really meaningful, trusting moment where she so appreciated me giving her this thing she really wanted. Jessie took the glove and carefully sniffed it. She then rolled it between her hands, making it into a little white ball. And then she ate it. Gone! I was, of course, horrified and embarrassed, and I had to go to the vets and tell them the stupid thing I had done. They told me everything would be okay and, happily, everything was. But it was a good lesson for me. I learned through all of these interactions that not only is each chimpanzee different in his or her approach to me, but that, when interacting with a different species, I cannot rely on human psychology. I had assumed Jessie was interested in the glove because she was curious about how it looked or felt or how it fit on her hand, but really, she wanted to know how it tasted. Definitely not something you would expect if you handed a glove to a human!

Each chimpanzee I have worked with has been different, but one of the joys for me is getting to know them as individuals.

GOOFY BONOBOS

As much as I am fond of chimpanzees, for all the reasons many people love them, I really fell in love with bonobos. Like chimpanzees, each bonobo is unique, with a special personality and demeanor. Kikongo is a bonobo I met many years ago at Lola ya Bonobo, a great sanctuary in the Democratic Republic of the Congo, where I do a lot of my work. As I mentioned, you can learn a lot about chimpanzees when you first meet them, but I am not sure I was prepared when I met my first bonobos. The contrast to chimpanzees was quite striking. Male chimpanzees are so cool and tough and they display and challenge you so directly. Of course, chimpanzees can be playful as well, but when I first started interacting with bonobos, especially young males like Kikongo, it became immediately clear that they are utter goofballs.

Kikongo is the silliest, funniest animal I have ever met in my entire life, and it was just impossible not to fall in love with him. Bonobos are known for making funny facial expressions that really do not fit clearly into an established list of behaviors, so no one really understands what they are supposed to mean. And Kikongo is the master of the funny face. Tongue wagging everywhere, lips contorted all over the place, and head shaking around. He would do flips and somersaults and try his best to get you to play with him. I remember sitting in front of him and literally just laughing and laughing—I couldn't believe what he was doing. He was so completely endearing and playful.

In this way, bonobos are sort of representative of why I fell in love with the animals I study. For instance, I really love wombats also and what they really have in common is that I just can't understand how these animals evolved to survive and be successful. How can an animal be that goofy and silly and make it in the world? Kikongo is the goofiest individual I know (human or nonhuman) and he has been tremendously successful. He was raised in a sanctuary but then was released into the wild and did very well out there. Bonobos tend to just represent what we as humans could be. Of course, chimpanzees are very capable of kindness, and they demonstrate these incredible friendships and caring interactions, but chimpanzees also have a very dark side. Contrasting chimpanzees and bonobos has always fascinated me given they are both equally related to human beings, so, in a way, they both represent the origins of our species. You know, there is not a single confirmed case of a bonobo killing another bonobo. Furthermore, and unlike chimpanzees where the males pretty much boss around the females, in bonobo society the females run the show, so there is really no infanticide or high levels of aggression that you see with chimpanzees, where there can be a lot of violence against the females, even toward mothers.

I am constantly trying to be a better bonobo. I realize they are not perfect, and they do show aggression at times, and they are not living in perfect peaceful communities up in the trees, but for me, they are a better model of what our species could be capable of.

VOICES AND PERSPECTIVES

Chimpanzees and bonobos face some pretty huge challenges to their survival. Many of the other contributors to this volume have touched on those issues: habitat loss, disease, the bushmeat trade, and the pet trade, which is often a by-product of the other issues. So, here, I prefer to consider what might be the start of some solutions and, in a way, since you are reading this, you are already part of the solution.

This book arose from a primatology conference held in Chicago in 2016: *Chimpanzees in Context*. One of the exciting things about this collection of essays, and the conference, is the seemingly simple initiative to bring people together. All sorts of different people working on different chimpanzee-related issues and studies come together to share ideas and their expertise—scientists and conservationists from universities, zoos, sanctuaries, and different field sites spread across Africa. The conference was one in a series of meetings that happen every ten years. These meetings are unique because there are really no other opportunities for people who have dedicated their lives to studying and protecting chimpanzees and bonobos to get together in one place and discuss the issues that these animals face and consider the strategies that can be, and have been, implemented to address them.

Richard Wrangham wrote a paper about the need for coalitions in conservation, and I think he has correctly identified the importance of coordinating efforts and working together. The second, and related, part of this challenge then is the need for a wider diversity of voices to be part of these discussions. We need to hear from, and listen to, voices from the countries where these apes are living. We need not only to hear from the senior experts in the field but also from students. And we need to be open to new voices from places like India and China, which both have a long history in primatology and, importantly, are themselves primate-range countries (though not apes, of course). How do we get people in some of the largest and fastest growing countries and economies in the world excited and interested in these conservation issues? I think this book represents the diversity of voices that can and should be part of all these discussions. We will need all of them if we are going to help save our closest living relatives on Earth.

14

RAVEN JACKSON-JEWETT

Dr. Raven Jackson-Jewett is the attending veterinarian at Chimp Haven, the largest chimpanzee sanctuary in the world. Nestled within two hundred forested acres in northern Louisiana, near the city of Shreveport, Chimp Haven is the forever home to more than three hundred chimpanzees, most of whom were once subjects of biomedical research in laboratories across the United States. In her role, she is charged with maintaining the health and well-being of the entire chimpanzee population at the sanctuary.

AN UNLIKELY INSPIRATION

I have always been a nurturer by nature, drawn to the sciences, and intrinsically motivated to help those who are incapable of defending themselves or having their own voice. Those who know me might describe me as a protective, compassionate caregiver. These characteristics make up who I am as an individual, and early on they quite naturally led to my interest in the field of medicine and my love affair with animals.

As a child I enjoyed my life growing up in New York City, the Big Apple. I lived in a high-rise apartment building and was lucky to be surrounded by constant love and adoration. One day, my mother and I walked hand in hand toward the noisy elevator waiting to whisk us down to the first floor as we always did when we set off somewhere. But on that particular day, when I stepped into the elevator, I came face to face with someone I will never forget.

He was dressed in a blue jacket, red shirt, jeans, and a baseball cap. He matched my height. The brown leather leash that was attached to him immediately grabbed my attention. I wondered why it was necessary, but then I focused my attention back to him and stared. I was fascinated with him, and I wanted to know more about him and get closer, but the more I tried the closer my mother held me. His bright eyes peered directly into mine as if he was as curious about me as I was of him. When we got off the elevator, I eagerly asked my mother, "What was that?" and she told me that he was a chimpanzee! That name was forever etched in my consciousness and that was the day I knew, somehow, I would have a closer connection with this species.

My initial encounter with that pet chimpanzee in a New York elevator led to my fervent explorations of wild chimpanzees. Inevitably, I stumbled upon Jane Goodall and her work with the wild populations in Tanzania. Even at a young age I admired her, calling her the "quiet storm" because I assumed she possessed an outward gentle spirit with a fiery heart nestled within. She seemed willing to withstand adversity for the sake of a completely different species and that was so admirable to me. My knowledge of her, coupled with my upbringing, inspired me to be the voice for those that couldn't advocate for themselves.

I became even further interested in chimpanzees after completing a wildlife and ecology course as part of my undergraduate curriculum. At the time, I was conducting behavioral observations of a chimpanzee troop housed at Montgomery Zoo in Alabama. There, I met John Paul, a juvenile chimpanzee whose mischievous nature caught my interest. His behavior and unique mannerisms were intriguing, but I desired to understand him on a deeper level. I felt a connection with him that at the time I couldn't explain. His eyes said, "There's more to me worth knowing."

It was the "more" in his eyes that reminded me of troubled children I have encountered who are funneled through various facilities due to behavioral issues but for whom no one ever stops to see what is at the core. Like these children, it seemed like the zoo visitors only focused on the chimpanzees' outward display of emotions, but I desired to understand the inward driving forces behind their behavioral tactics. It was increasingly clear to me that a chimpanzee is much more than just a physical being requiring medical care but one that is psychologically complex and emotionally advanced. Nothing would stop me from working with these amazing animals.

BUSTER AND ME

As the attending veterinarian at the world's largest chimpanzee sanctuary, I have been fortunate to meet many chimpanzees and build relationships with the chimpanzee residents for whom I care. Every relationship is a little different; however, one of particular importance to me is with an adult male chimpanzee named Buster.

My story with Buster actually began years before either of us got to Chimp Haven, when I was a student extern at a primate laboratory in New Mexico, which is where I got my first veterinary experience with chimpanzees. On my first day there, I was somewhat overwhelmed with the notable differences I was seeing between the research and zoological settings. My attention was immediately seized by a large male chimpanzee. His appearance was very distinctive: his face was much lighter and more mottled than the darker-hued chimpanzee faces I was accustomed to. But it wasn't

Buster

just his appearance that captured my enthusiasm to get to know him but the mischievous behaviors he displayed, not unlike John Paul from the zoo. I guess that is "my type"!

The staff quickly warned me about Buster's favorite thing to do, pelting them with well-aimed feces, and how it seemed to give him great joy at the expense of the target. I considered their warnings, but at the same time, I was very eager to know more about him. Each subsequent day, my visits with Buster and his group became my highlight. As soon as he would set eyes on me, he would seek out his large plastic barrel and begin to push it brusquely around the perimeter of his enclosure with much fervor and commotion. Of course, the intention of his display with the barrel was to ensure that I knew that he was the one in charge here, but each time he would abruptly stop and plop himself down in front of me. We would stay like that for a long time, simply staring intently into each other's eyes. When it was time for me to go, he would press his rotund belly up to the wire mesh for a quick friendly rub. These brief but meaningful daily interactions resulted in a lasting relationship between us.

One of the great aspects of this story is that we were reunited ten years later. In November 2015, the National Institutes of Health (NIH) announced that it would no longer support any biomedical research on chimpanzees and that all NIH-owned and NIH-supported chimpanzees would be retired to Chimp Haven. So, as the director of veterinary care and attending veterinarian at Chimp Haven, you can imagine my excitement when I was notified that we would be helping to relocate chimpanzees from the same research facility in New Mexico that housed the first chimpanzee to steal my heart, Buster.

I remember my heart beating rapidly as I scanned the master list of the chimpanzees housed at the facility in search of his name. One page down and fear set in as I did not see it. "Had he passed away?"

I asked myself. Cardiovascular disease is the leading cause of death in the chimpanzee population and Buster definitely fit the characteristics of young males who pass away unexpectedly due to sudden cardiac arrest. I felt the tears welling up, prematurely mourning the potential loss of my first chimpanzee love. But then, there was his name! On the day of his transport, I had the same feeling one has when anticipating seeing a loved one after a long period of time. When the truck pulled into the sanctuary, I felt my heart swell. He responded to my voice and, just like many years before, gestured for me to rub his belly. It was validation that chimpanzees are capable of building relationships that can last a lifetime. Buster and I have remained special friends, and ours is not unlike a close friendship between humans. We have our good days (sharing a secret hand gesture that no one else can decipher) and the occasional difficult one (grumpily not acknowledging one another when we have been offended). Buster and I are living proof that chimpanzees know the value of a long-term trusting bond; we have a friendship I will cherish for the rest of my life.

RECOGNIZING STRENGTH AND TRUTH

As an African-American female, I have faced many obstacles personally and professionally. Stereotypical clichés depict women of my ethnicity as exuding strength that does not correlate to the emotional pain they've endured at the hands of others. This has desensitized other ethnic groups from recognizing our truth. Oppressive measures taken against minorities have forced us to demonstrate a higher capacity of strength to overcome various obstacles. Society has taught us not only to exhibit strength but also resilience.

While actively providing hands-on care to chimpanzees I am constantly awed by their physical agility and strength. Observing a young male effortlessly toss a tire weighing forty-five kilograms across an enclosure quickly reminds you of chimpanzees' brute strength. However, it is not merely the physical strength of chimpanzees that encompasses who they are as a species but their spiritual resilience as well. As a medical professional, I can firmly attest to the chimpanzees' ability to recover rapidly from both physical injury and illness. Moreover, having knowledge of the research histories of many chimpanzees in the captive population for which I care, I can only imagine the psychological resilience necessary to recover from the invasive experiences they have experienced. Hundreds of chimpanzees have faced many extremely difficult research protocols for the purpose of advancing human medicine. It is truly amazing that they are able to build trusting relationships with the human population that also bears the face of their pain. I am honored to be a student of their astounding ability to endure insurmountable afflictions while continuing to exemplify strength and resilience.

My interactions with these chimpanzees have been an even greater influence on my personal ability to recognize that my strength is rooted in my willingness to walk in forgiveness and show love to those who have inflicted emotional pain on me. Although I can never forget the discrimination and social injustices that have occurred regarding minorities, I am that much more driven to be a productive steward to both humanity and the animal kingdom. Globally, the captive chimpanzee population has taught me to remain dedicated to my position in life and career by continuously exuding an attitude that promotes positivity and cohesiveness in a society that can appear to be solely focused on self-preservation. Fully aware of the sacrifices chimpanzees have made for the betterment of humanity, I am honored to practice sacrificial servitude in turn for them.

DREAMS, NIGHTMARES, AND SALVATION

Every day I arrive to work at Chimp Haven, it feels like I am walking in a dream. Forming and maintaining relationships with hundreds of chimpanzees who depend on me for their lifetime care is both daunting and immensely fulfilling. But, of course, there are always difficult days, both in terms of my professional life and how I balance my personal challenges as well.

Several years ago, I became the primary caretaker for my father as he suffered from end-stage disease. The requirements of my job make it extremely challenging to spend extended periods of time away from the sanctuary, so I struggled with finding the balance between being the primary caretaker for more than three hundred chimpanzees and the man who gave me life. Of course, I was able to take some time away from work to stay by my father's bedside, but even though I was only a phone call away, I worried about the chimpanzees all the time.

When my father eventually died, I was grief-stricken. Barely was I able to bury him before I, too, became gravely ill. I was diagnosed with a severe case of Stevens-Johnson syndrome (SJS) and toxic epidermal necrolysis (TEN) following an allergic reaction to an antibiotic injection. SJS/TEN leads to blistering and peeling of the skin and mucous membranes. Before I knew what was happening, I was lying helplessly in a hospital bed for the first time in my life. From the caregiver, I had become the patient.

Still grieving and experiencing excruciating pain, I was a complete mess. I was unable to speak and relied heavily on everyone around me to make medical decisions on my behalf. I remember thinking, "Is this what it feels like to have no voice?" I wondered if I would ever see the chimpanzees again. Even then, under those circumstances, I felt I was gaining a better understanding of what it was like to be voiceless and I vowed always to be the fiery advocate for the chimpanzees moving forward. Of course, my family was at the helm of my desire to regain strength and survive, but the animals that I care for daily have also left an imprint on my heart that makes them family, too.

Although doctors told me it would be about three months before I would recover and could return to home and work, I was released after just one. My first day back at the sanctuary was very challenging. I was nervous how the chimpanzees would react to me after being gone for so long. I was so happy that they were ecstatic to see me but also clearly saw they noticed something different about me. They immediately began looking me over and seemed to notice the subtle variations in my skin pigmentation. Many of them peered deeply in my eyes, staring intently at me, and I tried to let them know that I was okay. I was amazed that they could so astutely recognize changes in my physical appearance. Although I have worked with this population for many years, they always allow me the pleasure of learning something new about them.

INNOVATION AND THE FIGHT FOR LIFE

As a clinician, my primary focus is direct medical management of the chimpanzees in my care. One of my early challenges as a veterinarian was providing clinical oversight of a captive population of chimpanzees that had been experimentally infected with diseases that do not naturally occur in the species. This required a global view of the animals' health, including the potential of future unforeseen complications associated with past research, to develop a comprehensive medical plan in the best interest of each patient. I had to gain a vast understanding of the pharmacokinetics of therapeutic options and the potential benefits and risks of use of off-label medications in the chimpanzees.

Cotton's case was one that both challenged and inspired me to step outside of the box of pharmaceutical options for chimpanzees. He was born in April 1977 and arrived at Chimp Haven in November 2006. Cotton was known for his dramatic growls for treats and attention and the way he would present his entire bottom lip for his favorite juice. But his medical status was of particular interest to me as well. Cotton's research history involved experimental infection with simian immunodeficiency virus of chimpanzees (SIVcpz), the forbearer of HIV-1. I was able to determine from the research literature that chimpanzees with SIV not only suffer a negative impact on their health and reproduction but, in the wild, such chimpanzees can actually develop AIDS-like symptoms and can have up to a sixteen-fold increased risk of death. When I met Cotton, however, we knew virtually nothing about how to treat a chimpanzee experimentally infected with SIVcpz.

One's sense of urgency is impacted when you have the eyes of one you care about staring back at you. I knew Cotton's long-term prognosis was poor if I was unable to maintain his immune function and reduce his viral load. Ironically, my most advantageous innovation to date was the use of combination antiretroviral therapy that I gave him. This treatment was the product of biomedical research with chimpanzees that in turn was used to offer life-saving treatment in Cotton's case. Data from his case demonstrated that SIVcpz could cause immunodeficiency and other hallmarks of AIDS in captive chimpanzees but that combination antiretroviral therapy served as an effective therapeutic intervention. To this day, Cotton is the only chimpanzee in the world to have received this therapy, which enabled him to thrive for years at Chimp Haven. Later, we published a report on Cotton's treatment to ensure others could learn from our success.

FINDING PURPOSE

I entered into the veterinary profession desiring to specialize in chimpanzee medicine. I never imagined my dream would someday involve chimpanzees being retired from biomedical research during my tenure, and, furthermore, that I would be a key resource in relocation decisions and play an integral part in relocating hundreds of chimpanzees to their forever home.

I am most proud of the fact that I took the first steps toward realizing my dream of becoming a chimpanzee veterinarian. I did not allow the naysayers or my fears to deter me from what I was created to do. I believe that we are born with a key purpose to fulfill while on Earth, and our life experiences help us develop this purpose. In taking those immediate first steps in my own life, I placed myself in alignment for my part in the bigger picture of chimpanzee retirement from biomedical research. My life has been intertwined with that of the chimpanzee and I hope my legacy will demonstrate that nothing is impossible when you are in the proper position.

My advice to the next generation is to know the beauty behind why they were made—it is for a great purpose. We all have beautiful lives begging to be lived and, unfortunately, many people will never walk in that fullness because they will fail to embrace who they truly are. Our culture beats images of "success" into our brains daily. But truly accepting the magnitude of individual purpose begins with releasing misaligned ideas of self. Children should learn the importance of defining who they are at their core, standing firm in their personal values and vision. Success will forever be elusive if the world's standards become theirs. I challenge the next generation of veterinarians to look within themselves to determine what inspires them, to set their own goals, and to build authentic experiences, rather than conforming to others' expectations.

Life often requires us to move forward without seeing the entire plan, from a position of fear to faith. The best ideas in the world are lying dormant in someone's head. People should be inspired to invent a solution if it doesn't exist. I endorse evolving veterinarians to invest in the newer versions of themselves by becoming grounded in a community of people with whom they can create. This investment begins by giving themselves the freedom to make mistakes and embracing the unknown. There will forever be roadblocks between where one begins and where one desires to be. They should know that they will win at some things along the way and fail at others, but failures, too, are an integral part of the process of growth. My personal testimony is that all adversity faced continuously molds you into the person you are to become. Fear and self-doubt are natural, and life's encounters with perceived failures, heartache, and pain don't diminish you as a veterinarian because you have purpose. Fulfill that purpose.

15

FRANS DE WAAL

Dr. Frans de Waal is a Dutch primatologist who, for almost two decades, directed the research at the Living Links Center at Emory University in Atlanta. He is a best-selling author who has conducted research in a number of settings, including his seminal work on chimpanzee politics at Royal Burgers' Zoo in Arnhem, Netherlands, studies of bonobos at San Diego Zoo, and tests of chimpanzee cognition at the Yerkes Field Station near Atlanta, Georgia. His most recent book, Mama's Last Hug, examines the emotional world of animals and features the chimpanzee Mama.

COMPLEX CHIMPANZEES AND
PEACEFUL BONOBOS

Since childhood, I have always been interested in animals and animal behavior but not necessarily chimpanzees and bonobos. In retrospect, the fact that I much later ended up studying apes was mostly accidental. I could have ended up studying other species, birds or fish perhaps, but when I was a student I got involved in a psychology lab that had two young chimpanzees, and I found that I really liked working with them. Once I had studied chimpanzees, our closest relatives, everything changed. I was mightily impressed by their intelligence, which was of a different order than monkeys, dogs, or other mammals that I knew.

In the mid-1970s, I set out to observe chimpanzees at the zoo in Arnhem in my home country of the Netherlands. It was the first large, multimale chimpanzee zoo colony in the world. It was from those observations that I later wrote the book *Chimpanzee Politics*, in which I described the chimpanzee group's social dynamics. I was so impressed by the political intricacies among the chimpanzees that I decided I needed to stay with this species. Later, I also studied bonobos but, again, I was not necessarily driven by an interest in human evolution, even though the connection with humans was easily made.

In my work with the apes, I was most of all drawn to the fact that these animals seem to violate all the rules I had learned about animal behavior: animal behavior was supposedly either instinctive or explainable by simple associative learning. At that time, the lectures at the university took a very mechanistic turn, inspired by experiments on rats and pigeons run by scientists like B.F. Skinner. Inner processes (thinking, planning, consciousness, emotions) were denied, even ridiculed. It's not just that we could not know them, but who said that they even existed? When I watched chimpanzees, however, I decided that the simple rules I had been taught were clearly not applicable. I needed to know more.

The concepts of authority and hierarchy were viewed very negatively by my generation. This was the time of the student protest movement—hierarchy was bad, egalitarianism was good. Chimpanzees, however, live in a very hierarchical and power-driven system. The chimpanzee, therefore, represented a likeness of our predecessor, providing a picture that was quite the opposite of the idealized form that we students had for human society. The way in which male chimpanzees so overtly dominate female chimpanzees was not necessarily approved of by human society at large. Male chimpanzees often charge at females in the group and bully them with aggressive hitting and biting. Some studies suggest that the more dominant and aggressive male chimpanzees increase their chances of siring offspring. This hierarchical system was so obvious and overt that I started paying more attention to it in humans as well. I don't consider myself particularly power hungry and I tend to dislike the various human political games that are played. But more and more, after studying chimpanzees, I found myself noticing these interactions among people in my environment. It was even visible among the students, who all proclaimed themselves to be egalitarian and democratic. Sometimes their behavior was subtle but often it was more overt. In that sense, I think that watching chimpanzees has deeply changed my view of human society and the various political games we all play.

In *Chimpanzee Politics*, I began to compare ape behavior with Niccolò Machiavelli's *The Prince*. It was, of course, unusual for a biologist to consult the Florentine chronicler of human political intrigue, but it somehow made sense to me and subsequently inspired the label "Machiavellian intelligence" for social cognition in general. However, I consider this a rather narrow take because social cognition also includes the way mothers understand their offspring or how individuals manage to maintain peace despite social tensions. There is so much more to social intelligence than simply the struggle for power.

Later, when I studied bonobos, this realization created an enormous shift because bonobos are more egalitarian and peaceful than chimpanzees. Perhaps we share these traits with bonobos; despite the fact that we are very hierarchical, we tend to like egalitarian relationships. Many small-scale human societies have fairly egalitarian systems, and they work hard to maintain them. In this regard, therefore, the bonobos gave me a second and different view of our primate heritage, one in which the females are collectively dominant, and the society tends to be more peaceful than that of chimpanzees.

Maybe for this reason, many anthropologists do not like or acknowledge the bonobo. In their books about human evolution, they focus on chimpanzees and only mention bonobos in passing. They try to sideline bonobos as an "outlier" because they do not know how to make sense of female dominance and their peaceful societies. They assume that we humans got to where we are by means of violence and warfare, so where would bonobos fit in? They are perhaps also a bit embarrassed by the eroticism of the bonobo and so prefer to keep them on the sidelines. There are no good genetic or anatomical reasons, however, to favor the chimpanzee over the bonobo as a model of our shared ancestor. In my work, I have always defended the bonobo as an equally valid model of human ancestry and, if you do that, you will have a more complete picture. Considering both chimpanzees and bonobos offers a more relevant and accurate view of where we as humans came from.

CONFLICT AND RECONCILIATION

During my time observing the chimpanzees at the Arnhem zoo, I came to know each of them personally. Even though each individual was unique, one in particular had much more of an impact on me. His name was Luit, and his story is both dramatic and tragic.

He was an absolutely wonderful alpha male, one of the nicest I've known. Looking back, maybe he was too nice, because in the end he was tragically killed by two males in his group with whom he previously had a good relationship. To lose an individual I really liked and to be there in the last few minutes of his life was very hard on me. Not long before, I had begun studying conflict resolution and had discovered that chimpanzees reconcile after fights. At that time, people barely knew what to do with that finding as most research was about aggression and competition, winning and losing, and who gets the females, so to speak. So for me to say that chimpanzees reconcile after their sometimes-terrible fights puzzled many. People didn't know what to make of it and accused me of anthropomorphism.

When Luit was killed by the two other males, whom he knew very well, it was an eye-opener for me. Until that moment, I had always looked at reconciliation as something that was nice but perhaps not particularly essential in their lives. Luit's death showed me clearly that if the mechanism of reconciliation and peacemaking fails, then the consequences can be immense. It was a revelation. After that tragic day I decided I needed to know more not just about conflict and aggression but also about how aggression is kept under control and mediated. It led me to think carefully about the importance of peacemaking in society. It was also around the time of Luit's death that I decided to go to the United States to begin studying bonobos, who, I now know, demonstrate very different social mechanisms to deal with conflict. Luit's death was very sad for me, but it also put me on this new path to understanding the critical importance of how conflict is being managed.

HOW TO STUDY CHIMPANZEES?

I am often asked "Why do you keep studying chimpanzees and bonobos?" or "Don't we already know everything?" These questions don't come from fellow primatologists, of course, but from the general public. The truth is that we know almost nothing about apes compared to humans. Human behavior is being studied by two or three million scientists all over the world. If we bring all of the chimpanzee experts together, what do we have? Only several hundred people at best! We all know each other because there are so few experts in this field. It is clear to me that we have only scratched the surface of learning about chimpanzees and bonobos, and since these apes are under great threat in the wild, we sadly do not know how long they are going to be around for us to study and understand.

One of the key things I advocate for is to combine the efforts of those studying wild apes and those, like myself, who do captive work. Early on, I think many people thought of fieldwork as somewhat unscientific and inconclusive. Nowadays, fieldwork is more rigorous but there are still areas of study where it is insufficient and needs to be complemented by captive work. For example, based on field observations, you may claim that chimpanzees plan ahead and think about the future, but without a controlled experiment, it will be difficult to convince many people of this. Captive work remains especially relevant for claims about cognition and psychology. Let us say you are studying cultural transmission or imitation. Field scientists have observed chimpanzees cracking nuts with stones, and their work suggests that individuals learn from each other. But to know how they do so, and if it is based on imitation or some other mechanism, cannot be decided without experiments in a controlled setting. Experiments allow us to exclude many of the confounding things that influence findings from the wild, such as environmental influences or individual learning.

When the field and captive results fit together, the integrated result is more powerful than either one alone. In some cases, of course, the results will clash or lead to more questions. For example, we know that chimpanzees have a capacity for counting and quantity estimation, and that they can discriminate between different quantities presented on a computer screen. Then the question is whether they do that in the wild and how such skills help them to survive? One may discover things first in captive settings, and you then require field observations to obtain a full understanding of that behavior and its adaptive value. This is a very common path with chimpanzee and bonobo research, one that my own research has taken. I discovered that chimpanzees reconcile after aggression, such as by kissing and embracing each other. But some people argued at the time that perhaps they only do so in captivity because they are unable to get away from each other. Perhaps they won't do this in the wild where they have more space. But we now know from various field studies that reconciliations also occur in the wild, which in turn confirms the validity of what we see in captivity.

I think scientists studying chimpanzees in the field and captivity have a mutual respect now more than before and their work continues to influence each other in unexpected ways. We need both perspectives. If we had only fieldwork, we would have the problem of not knowing the precise mechanisms of some of the behaviors we observe. But if we had only captive work, without the context of their life in the wild, we would be unable to make evolutionary sense of their mental capacities. The fieldwork puts these capacities into an evolutionary and ecological context. Captive and fieldwork results are different pieces of the same puzzle. We need both.

THE CHALLENGE OF CHIMPANZEES

Studies with chimpanzees are quite often talked about in the popular media. The public is interested in the cognitive capacities demonstrated in experiments. But what many people do not realize is how hard it is to work with apes. You can run a study, for instance, about imitation (how chimpanzees learn from and copy one another), but it may take a long time to build a relationship with the chimpanzees that allows you to be successful and to refine the methods. We did one such experiment with chimpanzees at Yerkes and it took us over three years to get it right. Then my collaborators, Andy Whiten and Vicky Horner, went to do the same sort of experiment with children and they ran it in a couple of days! With children, you open the door, the parents come into the room with their child, you ask them to sit behind the table and tell them to "look at this, look at that," and, before you know it, you are doing the experiment.

With chimpanzees, it is not so easy. You need to develop friendships and trust with them in order to ask them to do things and participate in your studies. If you do not have that kind of close and friendly relationship, they may not do anything. They may be upset when you try to test them, which will not result in anything useful. Or they may be greatly distracted by their group's politics so they do not have time for your experiment because they want to get out and see their buddies. Or there may have been a fight in the morning and the chimpanzees are still stressed by it and will not pay attention to anything that you're doing. Basically, every day is a challenge to get the chimpanzees working with you under the right conditions and in the right mood for testing. This is why I am sometimes skeptical of negative findings in our field. The conclusion may be that apes cannot do this or that, and this may well be true, but before such a conclusion can be drawn, we need to make sure that our subjects were at ease in the experiment, understood the contingencies, were motivated to participate, and paid attention. If these conditions weren't met, you cannot conclude anything.

When I give public lectures, I will show a little video clip, one minute of an experiment, and, of course, the clip shows the best possible situation where the chimpanzee is doing the most interesting sort of thing. I do so to demonstrate the point I am making about the results of a study, but it does not reveal the realities of testing chimpanzees! It actually takes an enormous amount of preparatory work, trial and error, training, and relationship building with the animals before you get them to that point. The audience also comes up with ten different suggestions of experimental variations to ask new questions. These suggestions are always fascinating but typically not feasible. For example, testing a chimpanzee from one group with a chimpanzee from another group is virtually impossible because chimpanzees are xenophobic: you cannot test them with a stranger because they are very hostile to them. So even some seemingly simple proposals won't work, and methods that might work well with children cannot be translated directly to chimpanzees. This is what makes comparative research so challenging but also rewarding.

Furthermore, when you run studies with captive animals, you must think carefully about the conditions in which they are housed. This is important not only for the validity of the research, but also for the individual animals' welfare. Figuring out what is most important for chimpanzees is not simple and is often debated. I am on the board of Chimp Haven, the sanctuary for federally owned chimpanzees retired from labs in the United States, and I consider the habitats there to be ideal. They have large, forested yards and I don't see how things can get much better than that. The primary need of bonobos and chimpanzees, beyond the amount of space, is to live socially. Single housing is for me the worst thing you can imagine for highly social animals. So, for optimal welfare, they must all be housed socially, and in a large, safe, and complex habitat, one with space to escape groupmates that offers climbing opportunities.

PASSION, PERSEVERANCE, AND GOOD QUESTIONS

Forty years ago, when I was starting out on my career, many of my findings about chimpanzee and bonobo behavior were considered controversial. There was a time, not long ago, that if you were to say that chimpanzee reconciliation behavior was similar to human reconciliation behavior, people might object because you are implying that genetics is involved in the behavioral expressions of humans. But that resistance has largely evaporated over the years, which is a big relief. Comparisons with animals have become far less controversial than they used to be.

Now we can speak freely about animal intelligence, animal social behavior, animals' emotional repertoires, and broader parallels between animal and human intelligence and animal and human social behavior. People are fascinated by these concepts and are open to evolving ideas. I explored these topics in my book *Mama's Last Hug*. I think attitudes about animals and animal behavior research have dramatically changed for the better. I can only hope that such an attitudinal shift will benefit the animals themselves as well.

The field of primatology has always attracted people with dedication. You don't go into primatology to make money! It is not like a medical career or a career in law. Primatology is for those who really love animals, who like to work with them, and who want to understand them. That is also why primatological conferences are always fun because you can feel the love for the animals when talking to people about their study subjects. It is actually rare that you get a primatology lecture by someone who does not really seem very interested in the animals they work with. There is often a level of enthusiasm and dedication that, I think, is relatively rare in other fields. So, to be a successful primatologist, the first thing any student will need is that genuine love of being around animals. The work of primatology is simply too challenging to do it without passion and dedication. You cannot work day in and day out under the spartan conditions of the field or in a smelly lab if you don't have that sincere dedication.

Mama

Earlier, I described how hard it is to do the experiments that we do, which is another reason why you need people who really want to do those things. You have to find what you are genuinely interested in. Just being interested in the animals, or caring for animals, is not enough. You need more than that as your basis. You also need the intellectual curiosity and to have questions in mind. What are the issues that you're interested in? Research is continually revealing new, unanswered questions.

Technological advances are allowing us to ask new questions and in new ways. For example, at Emory University, we have a scientist who conducts noninvasive neuroscience on dogs because they are able to be trained. The dogs simply go into a neuroscanner and sit still for the few minutes to undergo a brain scan. We are not there yet with primates, although I envision that in ten years' time, we will have ways of doing that kind of thing. Noninvasive neuroscience is the future for cognitive research, and I think aspiring primatologists should look to take classes in neuroscience and prepare themselves for these new and innovative kinds of approaches to studying the primate mind.

16

ELIZABETH LONSDORF

The focus of Dr. Elizabeth Lonsdorf's career has been the study of chimpanzees living in Gombe National Park in Tanzania. Her research centers on the intersection of chimpanzee health and behavior and explorations of tool use and learning. She is currently an associate professor in the Department of Anthropology at Emory University and has previously held roles at Franklin and Marshall College and Lincoln Park Zoo, where she examined captive primate cognition complementing her study of wild chimpanzees.

A RAINY HIKE

I was not one of those kids who always knew I wanted to do this. I grew up with and loved animals and, of course, grew up watching nature documentaries. But early on, I never thought that it was something anyone else could really do. My sister was the one who wanted to be a veterinarian. She was into horses and other animals; I was into Star Wars. For a brief period (after a chance meeting with Carl Sagan), I wanted to be an astronomer and then an astronaut. Shifts in my interests in high school resulted in me enrolling at Duke University for my bachelor's degree, where I intended to major in either art history or pre-law. I was told from a young age that I was particularly good at arguing and that I would make a great lawyer if I managed to get through my childhood without my parents killing me!

Duke had several pre-orientation opportunities for incoming freshmen, one of which was Project Wild, an outdoor experience in the Pisgah National Forest. I was born around the Blue Ridge Mountains in North Carolina and thought it would be fun to get reacquainted with them while also getting to know my new classmates, so I signed up. We spent ten days hiking through Pisgah, carrying everything on our backs, making camp under a tarp each night, shifting back and forth between two sets of (progressively smellier) clothes. On top of that it rained for nine of the ten days. I remember being incredulous on the second morning of torrential rain when my crew leader told me to get packed up and ready to hike:

ME: "What?!!! We're still going to cover our miles today?!!"

HIM: "Uh, yeah, we hike in the rain."

And so we did, for that day and the next eight. It was glorious—I have never been so wet, cold, hungry, and dirty, but I loved it. I knew at that moment that I could never work in a law office or art museum all day every day; I needed to have some sort of job that allowed me time out to truly experience nature.

Once classes started at Duke, I was nearly thrown off the scientific track by miserably failing my first biology exam (I tell my students that to this day). Later on, though, I was lucky enough to take an animal behavior course from Peter Klopfer and a biopsychology course from Carl Erickson. Carl encouraged me to get involved with his research on the percussive foraging skills of the aye-aye, a unique and rare lemur species that uses an elongated middle finger to tap along tree branches to detect and extract burrowing larvae—delicious for an aye-aye. It was through discussions with my professors that I realized one could actually get a job doing such wonderful things. They were the ones who encouraged me to pursue a PhD, though at the time it seemed utterly ridiculous to me.

Then Jane Goodall came to speak at Duke for the thirtieth anniversary of the Duke University Primate Center (now known as the Duke Lemur Center). As I stood in line with my mother waiting to get my copy of National Geographic signed, I thought about what to say when it was my turn. Finally, it was my turn, and rather than saying something memorable and brilliant, I froze. I literally couldn't say anything except "thank you." My mother contends it was the only time she can remember me being struck speechless. Jane wrote "Follow your dreams" in my magazine, which I now know is a favorite autograph of hers, but it felt especially personal to me at that moment.

A year later, after working as an intern at the Epcot Center and Smithsonian's National Zoo, I interviewed at the University of Minnesota for a PhD in Ecology, Evolution, and Behavior. There I met Anne Pusey, who had worked under Jane Goodall earlier in her career. She told me about how she had been working to digitize the long-term data from the Gombe chimpanzee study and was looking for graduate students to work on the data. The opportunity to work on the massive Gombe data set was too incredible to pass up, even if it meant suffering arctic temperatures for the better part of six years living in Minnesota. I have been fortunate enough to work closely with Anne and Jane ever since, and I am still sometimes shocked at my good fortune to have those women and the Gombe chimpanzees in my life.

GETTING TO KNOW GREMLIN

Many of us studying chimpanzees are often asked about our favorite individual, but in my case, I read about this individual before

I ever met her. I first "met" Gremlin in Jane's books, in which she was described as the young daughter of Melissa. By the time I met her in person, she was a mother of four. When I first arrived at Gombe in the fall of 1998, Gremlin had just given birth to twin daughters, and miraculously they were both thriving (twins usually don't survive in the wild).

Gremlin was considered an expert at using tools to fish termites from their nests, which was the topic of my dissertation, so I knew I would be spending a lot of time with her. It is difficult to describe how wondrous it was to spend several days a month following Gremlin and her family and witnessing the twins continuing to thrive. She was an incredibly patient and solicitous mother, which I try to emulate in my own mothering (but often don't succeed). Twins are a burden for any wild chimpanzee mother, but Gremlin somehow managed. Anecdotally, I think this was in large part due to the twins' older sister, Gaia. She was always helping out with the twins, even attempting to carry them despite her own small size.

The twins' older brother, Galahad, interacted with the twins a lot as well, but he would often kidnap one of the twins in what I perceived to be an attempt to get his mother to stay with the larger social group when she wanted to go off on her own. She would patiently turn around, retrieve her baby from her son, and set off again down the trail. Galahad would lag behind, and Gremlin would sit and wait. Galahad would come to groom his mom, she would return the grooming, and then he would snag a twin and bolt back to the group. I saw this happen multiple times in a row until Gremlin seemingly lost her patience and took off down the trail without a second look back to see if Galahad was following. He eventually followed, seeming a bit disgruntled about it. At that time, the narrative in my head was that Gremlin was the consummate patient and caring mother and that if anyone could keep a set of twins alive, it was her. Indeed, the twins did survive and thrive and are mothers

Gremlin

themselves now, as is big sister Gaia, though big brother Galahad died of illness when the twins were still young.

When Gremlin became a grandmother, the chimpanzee I thought I knew suddenly showed a much different pattern of behavior. To the surprise of everyone, she began kidnapping her grandchildren! Emily Wroblewski published the first description of such behavior, when Gremlin took Gaia's firstborn infant. Emily presented the idea that Gremlin may have been motivated to take the baby as a protective measure due to the arrival on the scene of another female who had previously attempted to attack Gremlin's newborn babies. Since then, however, Gremlin has done this two more times, even kidnapping a set of twins Gaia gave birth to (twins seem to run in this family). Work is ongoing to make sense of these incidents, but, to me, they are a great example of how chimpanzees always surprise you. It seems that she has moved past this behavior now, but you just never know.

SONS AND DAUGHTERS

What motivates me to study chimpanzees? The answer to this is quite simple—I love doing it. I suppose the more scholarly answer is that I am driven to create new knowledge about one of our closest living relatives, which is certainly true as well. But the more essential motivation comes from the fact that working with chimpanzees, and the people who study them, is flat-out fun and fascinating. More than with any other animal, I see us reflected in them. Dominance hierarchies, alliances, the rise and fall of alpha males, kids tussling, and queen bees; all of these are present in any group of chimpanzees or humans.

I remember quite clearly the first time I met Fifi, who was perhaps the most well-known wild animal in the world. Fifi and her family were made famous by Jane's books, *National Geographic* articles, and documentary films. By the time I met her, it was like meeting a rock star. She was lying down in the forest of Gombe, resting with her newest daughter, Flirt, and soon proceeded to perform a classic F-family behavior: dangling her baby by her feet and tickling her in the chimpanzee version of "airplane." That was in 1998, and I have spent over twenty years since studying chimpanzee infant development. I have been lucky enough to watch an entire cohort of young females grow up and become mothers themselves. I have watched rambunctious young male babies ascend to alpha male status and become infanticidal leaders, while others have grown up to be gentle, lower-ranking members of the community.

During this time, I was myself learning the rules, alliances, and hierarchies of graduate school, post-PhD employment, and establishing a professional space for myself. I was also navigating motherhood—first a daughter and then a son. When they were babies, and still now, I think about which chimpanzee mom I should emulate. Should I be more like Fifi, who was rather laissez-faire but ferociously protective when the situation called for it? Or should I be like Gremlin, a solicitous and protective mother who keeps her kids (and grandkids) quite close? I certainly don't want to be like Patti, a very distant mother who seemed incapable of attending to more than one offspring at a time.

I also continually notice distinct behavioral differences in my two children. My son lives life in a constant state of rough and tumble play, as if he were fueled with five gallons of espresso (he isn't), while my daughter is quieter and more mellow. My son had two visits to the emergency room before the age of three; my daughter has yet to have one (knock on wood). They are the quintessential two sides of the risk-prone/risk-averse coin, a sex difference that many in our human society would chalk up solely to gender socialization. However, I have spent the past several years investigating sex differences in young chimpanzees, and there are very clear parallels to the differences I see in my own children, but in the absence of intensive cultural gender socialization. The nature/nurture dichotomy is a false one, of course, and sex differences occur on a continuum, but it is fascinating to see such similarities in chimpanzee and human children. These parallels, and the ever-changing soap opera of the Gombe chimpanzee communities, are the fundamental motivation for what I do.

Interestingly, these themes of sex differences played out very early in my career as well. My first scientific publication was a brief communication in *Nature* that described sex differences in learning of a tool-use skill in the Gombe chimpanzees. In it, my coauthors and I presented data that described how young female chimpanzees, on average, develop a tool-use skill known as termite fishing much earlier than young males. Termite fishing involves selecting a lengthy piece of vegetation, stripping off any lateral branches, and inserting the resultant tool into a hole on a termite mound. The termites inside the mound attack the tool as an intruder, and then the chimpanzee carefully extracts the tool and eats the clinging termites. Immature chimpanzees learn this skill by close observation of their mothers and subsequent trial-and-error practice on their own. My coauthors and I showed that daughters spent more time watching their mothers, and sons spent more time simply playing at the termite mound. While daughters were often closely attending to the task, sons were doing somersaults, annoying other individuals who were termite-fishing in attempts to get them to play, and generally goofing off. As you might expect, daughters acquired this skill earlier, were more efficient at acquiring termites at younger ages, and matched the termite-fishing techniques of their mothers. Sons acquired the skill later, were not as efficient, and mostly converged on a simpler termite-fishing technique. These were exciting findings, both from the perspective of sex differences and documenting social transmission of tool-use techniques in the wild. As a newly minted PhD, I was ecstatic when it was accepted for publication in a prestigious journal.

Once published, the article garnered the attention of the news media and, as often happens, some headlines and stories got it "right," while others embellished the conclusions into a "boys are dumb" narrative: I was faced with headlines such as "Boy Chimps Chumps Just Like Us" and "Boy Chimps Are Lazier Than Girl Chimps." The most surprising response was when a "Men's Rights" website excoriated the findings as both frivolous and biased because "all of the authors are women." My coauthors were my adviser, Anne Pusey, and my statistical mentor, Lynn Eberly. While the other incorrect or embellished headlines were annoying, this interpretation really ticked me off. They were not only putting down the research as unimportant; they were insinuating that the only reason we had found that females were better at something was because we were females ourselves.

RESEARCH INNOVATIONS

It's funny to look back and think of conducting detailed frame-by-frame video analysis as an innovation because now it is an essential technique for studying animal behavior, either for primary data collection or as backup. However, when I started my dissertation research on the topic of tool-use development in chimpanzees, very little had been done to study this question in wild chimpanzees let alone by employing frame-by-frame video analysis to study it.

At the time, the logistics of keeping a video camera (Hi8, then miniDV) functioning and powered in the field, especially in the rainy season, were not trivial. I spent a very overwhelming day running around BP Solar in Dar es Salaam, Tanzania, buying my own solar panel, its wiring and fittings, and a gigantic truck battery, all of which I hauled across the country on a two-day train journey. Once I got to Gombe, Anthony Collins gamely helped me install the contraption on the roof of the house where I would be staying, for which I'm eternally grateful because I had no idea what I was doing. After a few hiccups, I had a fully powered, handheld video camera with which to collect hours and hours of data, which I then had to spend even more hours analyzing.

My refrain became that video data is a blessing and a curse: you miss very little, but there is always more that can be analyzed (especially from the point of view of a dissertation committee!). As it happens, I am still using those videos more than two decades later, most recently to conduct a comparative analysis of termite-fishing by chimpanzees at the Goualougo Triangle Ape Project in the Republic of the Congo, in collaboration with Crickette Sanz and David Morgan.

A second innovation most pertinent to my work is all the amazing things you can find and analyze in chimpanzee poop. When I first started working at Gombe, I never imagined that I would one day spend ten nail-biting days, as I recently did, tracking a box of chimpanzee poop from Tanzania to my collaborators in Illinois. Oh, the wonders that are extractable from these noninvasively collected gold mines! Fecal samples can be analyzed to determine dietary intake, paternity, parasite and bacterial infection, viral status, reproductive hormone levels, stress levels, immune status, and weaning stage, among other things. At Gombe, the research team has used fecal samples to track transmission and pathogenicity of simian immunodeficiency virus, determine how differing alpha male strategies do, or do not, result in fathering offspring, and understand how female dominance status relates to physiological stress, just to name a few. New noninvasive diagnostic tests for pathogens are being developed every year, which will afford us even more opportunities in the future to track and mitigate disease risks. I never thought I would say it, but to this chimpanzee researcher, a fecal sample is worth its weight in gold.

ADVICE TO ASPIRING PRIMATOLOGISTS

I have been fortunate to receive mentorship from some incredible primatologists, and I would like to end my reflections here, with thoughts and ideas that I hope may be of help to others as they become started in their careers.

In my opinion, the most important thing that aspiring primatologists can do is find a way to get your hands dirty. Either literally, by volunteering in a lab or at a field site, or figuratively, by entering data, coding video, or collecting behavioral observations, etc. Find a way to get scientific experience. It doesn't even have to be with primates—one of the world's experts on chimpanzee welfare, Steve Ross, started out studying pigs! Just find a way to make a meaningful contribution to a scientific research effort so that you can start building both your skills and evidence of your dedication. In reality, much of a primatologist's life is spent doing things like data entry, data analysis, lab procedures, grant writing, teaching, etc., so look for opportunities to do any of those things.

Second, if possible, don't restrict yourself in terms of geographic areas, a particular species, or a particular topic; take what opportunities you can get. I would never have chosen to spend six years in Minnesota given how much I hate the cold, but that is where the graduate lab that I wanted to be a part of was located, so I did. Where you start and where you end up are quite likely not going to be the same place, so don't restrict yourself unnecessarily. Likewise, don't restrict yourself in terms of the skills that might help your research. Get over any fears you may have about math, technology, computer programming, or lab work—all of these are critical skills for current and future primatologists and will open larger avenues of inquiry to you.

Third, build your scholarly network at every opportunity: reach out to scientists whose work you admire, organize a symposium for a conference on your topic of interest, seek out opportunities for collaboration, and attend on-campus seminars and stay afterward to talk to the speaker. Chimpanzees are incredibly social animals, so channel your inner chimpanzee and get out there and be an active participant in the field of primatology.

AFTERWORD

LYDIA HOPPER

Dr. Lydia Hopper is a comparative psychologist who studies how primates innovate and learn new skills. Through her career she has been fortunate to work with chimpanzees in zoos, sanctuaries, and research settings, and she is dedicated to applying what she learns about primate cognition and behavior to enhance the welfare of captive monkeys and apes. She is the coeditor of Chimpanzee Memoirs *and* Chimpanzees in Context. *When working on this book with Stephen Ross, she was the assistant director of the Lester E. Fisher Center for the Study and Conservation of Apes at Lincoln Park Zoo, Chicago, but has since become an associate professor at the Johns Hopkins University School of Medicine in Baltimore.*

I am so thankful to our contributors for their generosity in sharing their personal experiences through their essays in this book. It was my honor to read their stories. As a scientist, I am fascinated by "how" and "why" questions, and the essays in this volume provide countless illustrations of how and why primatologists enter into their careers—what motivates them, how they got to where they are today, and what lessons they've learned along the way. These are not stories that we often get to hear. We typically only get glimpses through published articles or curated talks.

The "finished piece" is fascinating, but it is just part of the story—the metaphorical tip of the iceberg. Learning about what inspired a primatologist in the first place, the challenges they faced along the way, and how they tackled those challenges really gives us a behind-the-scenes glimpse into the lives of these "chimpologists."

My own work has centered on studying the cognition and welfare of captive primates. My interest in primates was sparked as a young child when I was fortunate to observe orangutans playing in the jungles of the Sepilok Rehabilitation Center in Malaysia. Later I observed the behavior of orangutans who lived at Chester Zoo while an undergraduate student at the University of Liverpool. But my deep interest in primate cognition was forged during my time studying for my PhD with Andy Whiten. I conducted my studies of chimpanzee culture at a research facility in Bastrop, Texas. There, I not only found a deep appreciation for chimpanzees' social and technical intelligence, but I was also fortunate to learn from the experts there about the ways in which we can refine and enhance the way we care for captive primates.

A real joy for me in my work is getting to know the individual animals—to learn their personalities, to gain their trust, and to watch as they make new discoveries. I have had the great fortune of working with many chimpanzees over the years and so, for me, it is impossible to pick a favorite. I have been delighted equally by the curious problem solvers, the gentle peacemakers, and the manipulative socialites. I am always fascinated to see how each individual responds differently to the cognitive puzzles that I present them with as part of my research.

I study how chimpanzees innovate, how they learn from one another, and who they choose to copy. In running one study, I was interested to see if chimpanzees, like humans, would copy a high-ranking, prestigious individual even if it meant going against their own preferences. Specifically, I trained a dominant female chimpanzee, Josie, to exchange red plastic tokens for pieces of carrot. The rest of her group followed suit, exchanging the red tokens with me for carrot pieces, in spite of a better option also being available: white tokens that could be exchanged for much tastier grapes. I have a video clip from this experiment that I often share in talks. It shows the chimpanzees in Josie's group lining up at the perimeter of their enclosure to exchange red tokens with me, even though they then discard the unwanted carrots they receive in return. However, just to the right of the group, you can see a young male chimpanzee eagerly bobbing up and down. He is trying to initiate play with me. Meanwhile, I am trying to be an impartial scientist and conduct my study. But it is tricky because he is so persistent and so playful. And this is how nearly every test session went with Krog. He just wanted to play all the time! When I show the video in talks, to this day his antics make me giggle. His happiness to see me and his joyful participation in my study on his own terms is a perfect example of why I love my job and why I continue to work to learn more about chimpanzees so that we may better understand them and better care for them.

Reading the essays in this book was a real treat for me, and I hope for you, too. I gained new insights into my colleagues' careers and learned lessons that will inform my future work. I hope that this volume also provides inspiration to others. Of course, everyone's story is unique, shaped by time and place, as well as the path that each person chooses to forge. It was for that reason that my coeditor Steve and I wanted to share the stories of a diverse group of primatologists, including those whose careers may be less well known to readers outside the field of primatology. It was important for us to present a diversity of perspectives, across professions and interests, and, most importantly, to share the experiences of primatologists who grew up in the countries where chimpanzees live in the wild. However, in spite of the varied career paths shared here

Krog

by our contributors, a universal theme runs through their essays: everyone's clear passion for their work. The deep connection our contributors have made with the chimpanzees they study and their continued dedication to understanding them, caring for them, and conserving them, are common narratives throughout this book.

As we look to the future, we must work hard to safeguard one for wild chimpanzees. As echoed throughout this volume, chimpanzees, like many primate species, face a precarious future due to the myriad of threats they face from human activity. Equally important is to ensure the well-being of those chimpanzees that rely on human care in captive settings. Success in either sphere requires communication—communities coming together to share expertise and insights, to listen to each other's perspectives, and to collaborate in their efforts. This can only be achieved if we prioritize strategies to recognize, respect, and promote the work and knowledge of local communities, both in our future endeavors as well as when presenting historical work. Additionally, the field of primatology must create and maintain an environment that is accessible, safe, and welcoming for all. In doing so, the field will not only be able to better serve those who wish to engage with primatology, but the science will be richer for a diversity of perspectives and expertise.

Donna Haraway noted that "stories are technologies for primate embodiment," and in this book we have shared the stories of many prominent primatologists whose careers have orbited around our chimpanzee cousins. I hope that this volume is not the last such compendium and I look forward to reading about the careers, achievements, and learnings of future chimpologists whose careers are just starting or have not even begun at this time.

SUGGESTED READING

Among the many common themes that weave throughout the essays in this collection is the inspirational power of books. Over and over, these chimpanzee experts have recounted how reading has opened their eyes to new information and novel perspectives that ultimately led them to their career paths. As such, we provide here a list of the books mentioned throughout this volume and encourage you, the reader, to find your own inspiration within their pages.

Adamson, Joy. *Born Free: A Lioness of Two Worlds.* New York, NY: Pantheon Books, 1960.

Boulle, Pierre. *Planet of the Apes.* New York: Random House, 2011.

Burroughs, Edgar Rice. *Tarzan of the Apes.* Racine, WI: Whitman, 1964.

De Waal, Frans. *Chimpanzee Politics: Power and Sex Among Apes.* Baltimore, MD: Johns Hopkins University Press, 2007.

De Waal, Frans. *Mama's Last Hug: Animal Emotions and What They Tell Us About Ourselves.* New York: Norton, 2019.

Goodall, Jane. *The Chimpanzees of Gombe: Patterns of Behavior.* Boston: Belknap Press of Harvard University Press, 1986.

Haraway, Donna J. *Primate Visions: Gender, Race, and Nature in the World of Modern Science.* New York: Routledge, 1989.

Hare, Brian, and Woods, Vanessa. *Survival of the Friendliest: Understanding Our Origins and Rediscovering Our Common Humanity.* New York, NY: Random House, 2020.

Heltne, Paul G., and Linda A. Marquardt. *Understanding Chimpanzees.* Cambridge, MA: Harvard University Press, 1989.

Hopper, Lydia M., and Stephen R. Ross, eds. *Chimpanzees in Context: A Comparative Perspective on Chimpanzee Behavior, Cognition, Conservation, and Welfare.* Chicago: University of Chicago Press, 2020.

Lawick, Hugo V., and Jane Goodall. *Innocent Killers.* London: Collins, 1971.

Lonsdorf, Elizabeth V., Stephen R. Ross, and Tetsuro Matsuzawa, eds. *The Mind of the Chimpanzee: Ecological and Experimental Perspectives.* Chicago: University of Chicago Press, 2010.

Lorenz, Konrad. *King Solomon's Ring.* New York: Routledge, 2003.

Marler, Peter R. *The Marvels of Animal Behavior.* Washington, DC: National Geographic, 1972.

Schaller, George B. *The Year of the Gorilla.* Chicago: University of Chicago Press, 1964.

Wrangham, Richard W., William C. McGrew, and Frans De Waal, eds. *Chimpanzee Cultures.* Cambridge, MA: Harvard University Press, 1996.

Wrangham, Richard W., and Dale Peterson. *Demonic Males: Apes and the Origins of Human Violence.* Boston: Houghton Mifflin Harcourt, 1996.

CPSIA information can be obtained
at www.ICGtesting.com
Printed in the USA
JSHW021529260622
27275JS00004B/4